Successful Professional Reviews for Civil Engineering Technicians

emerald
PUBLISHING

ice
Publishing

Successful Professional Reviews for Civil Engineering Technicians

Malcolm Peake

Published by Emerald Publishing Limited, Floor 5, Northspring, 21-23 Wellington Street, Leeds LS1 4DL.

ICE Publishing is an imprint of Emerald Publishing Limited

Other ICE Publishing titles:
Civil Engineering Procedure
Institution of Civil Engineers. ISBN 9780727764270

Successful Professional Reviews for Civil Engineers, Fifth Edition
Patrick Waterhouse. ISBN 9780727766090

Mentoring for Civil Engineers
Patrick Waterhouse. ISBN 9780727764300

Inital Professional Development for Civil Engineers, Second Edition
Patrick Waterhouse. ISBN 978-0-7277-6098-2

A catalogue record for this book is available from the British Library

ISBN 978-1-83549-943-6

© Malcolm Peake 2025

publishing under exclusive licence by Emerald Publishing

Permission to use the ICE Publishing logo and ICE name is granted under licence to Emerald from the Institution of Civil Engineers. The Institution of Civil Engineers has not approved or endorsed any of the content herein.

Cover photo: Lebedovskaya/stock.adobe.com
Commissioning Editor: Michael Fenton
Content Development Editor: Cathy Sellars
Production Editor: Benn Linfield
Typeset by KnowledgeWorks Global Ltd.
Index created by David Gaskell

Contents

Preface

The Institution of Civil Engineers (ICE) requirements for professional qualification are set out in its guidance documents:

- *Technician Professional Review Guidance*
- *Civil Engineering Technician Apprenticeship Level 3 Version 1.1 End Point Assessment Guidance*
- *Civil Engineering Senior Technician Apprenticeship Level 4 Version 1.1 End Point Assessment Guidance.*

As a licensee of the Engineering Council, the ICE is required to follow the *UK Standards for Professional Engineering Competence and Commitment, 4th edition (UK-SPEC)*. This sets out the prescribed standards for competence and commitment required for registration as EngTech.

The Engineering Council have also published *The Approval and Accreditation of Qualifications and Apprenticeships (AAQA), 1st edition.* This sets out the policy, context, rules and procedures for recognising learning and development programmes that help develop competence and commitment to the UK-SPEC standard.

The Attributes (or KSBs) to be demonstrated to the ICE at the conclusion of a trainee's initial period of development reflect those standards demanded by the Engineering Council. This book is written to assist trainee technicians, apprentices and those responsible for their development.

Acknowledgments

This book has been a long time in the making. After several years of anticipation and dedication, it has finally come to life. The journey has been long and filled with countless challenges, difficulties and many moments of doubt. I am deeply grateful to everyone who has supported the production of this book. Your encouragement and belief in this project have been the driving force behind its completion.

First and foremost, I would like to thank my family for their unwavering support and patience. You have given me the time and space to dedicate myself to bringing this book to life. In addition, I have been fortunate to be able to draw on the help, support and inspiration from an experienced panel of people from across the industry. This support has been invaluable in helping to ensure this book reflects the broad opinions of the industry. I extend my thanks to you all for your ever-present support and the countless advice, anecdotes and challenges you have all provided in the writing of this book. Your insights and enthusiasm have been invaluable.

With this in mind, I would like to offer my thanks and recognition to Ayo Ajayi, Sam Ashman, Steve Byrom, Laura Clegg, Simon Dunbar, Rob Ehren, Jayne Geary, Kate Harrison, Alex Hudson, Steve Hyde, Ibrahim Kapasi, Sebastian Kreft, Jill Mascarenhas, Sean Melody, Jason Nell, Philip Parker, Aaron Passfield, Tom Player, Amy Pledge, Tom Price, Mike Rogers, Hazel Sanders, Jason Shannon, Hannah Shewan-Friend, Annabel Smith, Rich Tournay, Ingrid Valceschini, Jean Venables and George Woods.

Writing this book has been an amazing journey. Special thanks go to Mike Clark (you are an exceptional communicator), Sue Hawkes (your poem is amazing, your passion and drive to see real change for technicians are limitless), Tim Brownbridge (your career reflects a deep commitment to advancing the construction industry through education, training and professional development), Katherine Warner (you have helped me navigate the intricacies of apprenticeships) and to my wife Jo Peake (your attention to detail and strong grammar skills are second to none, your support and patience have been invaluable). I am also immensely grateful to the editors, Cathy Sellars and Michael Fenton, whose keen eye and thoughtful feedback coupled with a consistent enthusiasm have helped shape this book into what it is today.

This book has been inspired by a similar set of books for engineers written by Patrick Waterhouse, who combines a natural enthusiasm for professional development and a deep well of knowledge with a relaxed and engaging communication style. I am grateful for all of Patrick's advice and guidance in helping me to gain the confidence to get this book started. I hope I have produced something that is worthy enough to complement the respected range of books written by him, Mac Steels and Jean Venables.

Although I have benefited from the considerable wisdom, advice and support from a large panel of people, I take full responsibility for any mistakes contained in this book.

Malcolm Peake
November 2024

Prelude

I met Susan Hawkes when she was 'just' doing a bit of drawing. Neither of us had any idea what would happen next. Sue not only became a technician member of the ICE, she volunteered as the local branch rep, followed by vice-chair and then chair. Soon afterwards she stood for the ICE Council. She is currently working with the Engineering Council on their vision for change.

Sue is currently a Reviewer, End Point Assessor and Fellow of the ICE.

Shortly after passing her technician professional review, she penned this poem. I am including it here for you all to enjoy.

Hello, my name is Sue, I'm a CAD technician, how do you do?
Every day, I draw, every day, a little bit more.
I drink coffee, and I drink tea, I talk to people, and they talk to me.
At the end of the month, they give me money, it's called a salary, ain't that funny?

I'm an Eng Tech MICE, letters after my name, it makes no difference coz I'm still the same.
But the degree of respect I get from my peers, truly amazing, brings me to tears.
I go to work, I do my bit, all day long, I draw and sit.
Don't get me wrong, I do work hard, I play my role, I do my part.
I'm part of a team, a wheel in a cog, we all work together, we all do our job.

The design team give me the info I need, I produce my drawings for the client to read.
They give their approval, say it's alright, then it goes to the pilers who take it to site.
They build the cofferdams and walls of sheet piles, then my drawings get put into their files.

The client then pays us for a job well done, or at least that's how they say the company's run.
I've been on a course to learn about cranes, but I soon discovered I had to use my brain.
Appointed Person they said I'd be, but I drew the line at the responsibility.

I'd get the knowledge, the info I need, and leave it at that, I said I'd concede.
I'm not an engineer, never wanted to be, I just wanted to draw and to be happy.

Now I get paid for doing a job like this, I really love it, it's bliss.
I've joined a committee part of the ICE, to make a difference, but, I'm still me.
I organise events for learning and fun, and if you get the chance you should book on and come.
A STEM ambassador I've also become, to show young people that engineering is fun.
So, don't forget my name, its Sue, and I'm a CAD technician, thank you.

Sue Hawkes EngTech FICE FWES

About the author

Malcolm Peake is a professionally qualified chartered civil engineer and a Fellow of the Institution of Civil Engineers (ICE). He has extensive experience in the design, management and delivery of various road, rail and marine engineering projects. Over his career, he has changed his focus from developing solutions to solve civil engineering problems to one helping people to develop their professional careers.

As a membership development officer, he worked for a decade directly for the ICE, developing and implementing initiatives while providing advice and guidance to aspiring members. Since then, he has moved to Tony Gee and Partners where he supports the company's ongoing focus on learning and development while embracing its unique culture. This includes managing the initial professional development of students, apprentices and graduates as they build the foundations for their own careers.

Malcolm has always sought to engage, share and learn about professional development. He is a member of the Civil Engineering Employers' Training Group (CEETG), vice-chair of membership for ICE South East England Regional Committee and a member of the ICE Kent & East Sussex branch. He has been a Reviewer for over 10 years and an End Point Assessor since 2020 and a trainer of new reviewers.

Introduction

There are several books and guides aimed at supporting engineers who wish to forge a career in civil engineering. There are some well-known, excellent books: *Initial Professional Development for Civil Engineers*, *Mentoring for Civil Engineers* and *Successful Professional Reviews for Civil Engineers*, all authored by Patrick Waterhouse and published by ICE Publishing. But so far there has not been something similar for civil engineering technicians. This book has been written to fill that gap.

This book is for people who can solve practical engineering problems in the civil engineering industry. The aim of the book is to provide a useful guide to show how it is possible to use those skills and experiences to become professionally recognised as a civil engineering technician with the Institution of Civil Engineers.

This book is primarily focused on the ICE reviewing process for people applying directly for recognition as EngTech MICE or using the apprenticeship route. It is an accumulation of years of experience from a diverse pool of knowledge across the civil engineering industry. By reading this book you will gain an understanding of what you need to do to apply and be successful in either your technician professional review or your end point assessment.

REFERENCE

EC (Engineering Council) (2024) *The EngTech eBook*. Engineering Council, London, UK. https://www.engc.org.uk/engcdocuments/internet/WebsiteEngTech20eBook.pdf (accessed 05/09/2024).

Malcolm Peake
ISBN 978-1-83549-943-6
https://doi.org/10.1108/978-1-83549-940-520251001

Chapter 1
The history of technicians

This book aims to provide a practical guide on how to approach a professional review, offering step-by-step instructions, tips and best practice. Whether you are a seasoned professional or new to the field, this book will give you the knowledge you need to prepare and succeed in your review. However, before diving into the details, it is worth stepping back and briefly looking at the history of technicians.

Alternatively, if you are the type of person who is not great at reading lots of text or simply doesn't have time for unnecessary details and just wants to get things done, skip this chapter and head directly to chapter 5 where you will find some clear steps outlining how to achieve your goals.

Being a member of a professional organisation means you are part of a community of professionals who share similar standards, ethics and goals. It's about sharing knowledge, supporting each other and upholding high standards in your work. It could be worth getting to know the Institution of Civil Engineers to get a sense of whether its values align with your own before taking the plunge and applying to join. This chapter is intended to provide a historical context, demonstrating how technicians have been the backbone of the industry and giving an insight into the evolution of the professional standards and practices.

Civil engineering technicians have always played a vital role in the civil engineering industry. From the very beginning technicians have had key roles in the design and delivery of engineering solutions. They have been an integral part of the industry, translating theoretical knowledge into practical solutions, shaping the built environment of their time.

However, their contribution to the civil engineering industry has often gone unrecognised. For most of the history of science and engineering, technicians have been invisible. Technicians are often the forgotten people. If acknowledged at all, they have been referred to as CAD monkeys, tracers or lab rats. Part of their role involved giving up their independence and giving up on any credit, working instead as nothing more than an assistant to the engineer (University of Bristol, 2023).

The Industrial Revolution

Things started to change during the Industrial Revolution. This arguably changed the whole of civilised society. Beginning in the UK, many technological innovations started to be developed during the late eighteenth century and continued into the nineteenth century. This marked a significant turning point in history. The Industrial Revolution influenced almost every aspect of daily life. The mechanisation of industries and the growth in steam power brought about new ways of working and living, resulting in a surge in demand for new technical skills.

Technical skills were developed across many areas, starting in the agricultural and textile industries through the mechanisation of power and the invention of new machinery. This industrialisation was boosted by the development, design and construction of ports, canals and roads to help move raw materials and finished products more efficiently. This increased the need for resources and subsequently fuelled and inspired further demands for industrial progress in mining and the generation of new sources of energy. The invention of steam trains and growth of the rail industry continued the need to develop technically capable people. This is also recognisable today as we transition from a digital revolution to a technological revolution.

Did you know?

Industrialisation required people to develop a knowledge and understanding of engineering. Since the eighteenth century there has been an increased need to have people capable of applying proven techniques to solve practical engineering problems. Those problems could be in a variety of situations, including the design, manufacture, maintenance or demolition of various products, equipment, processes or services.

Not only did the Industrial Revolution drive the need to develop technically capable people, but it also brought about a new organisation of work known as the factory system. This required improvements in the need to organise resources to ensure work was done with consideration to cost, quality and safety. This created a need to have managers, foremen and others capable of organising the delivery of work. The factory system enabled the manufacture of goods to consistent, precise specifications. This enabled work to be completed reliably and by following quality processes.

During this time the location of work changed from rural areas and concentrated workers in cities and towns. The increased urbanisation of the population initially resulted in crowded, substandard housing and poor sanitary conditions. The Building Act 1774 drove improvements in the construction of housing, while in 1865 Bazalgette's sewers industrialised the water sanitation industry. The Industrial Revolution also brought about a desire to be ethical alongside the regulation of health and safety.

Did you know?

The first modern piece of health and safety legislation was the Morals of Apprentices Act 1802, which was the first of hundreds of pieces of regulation that led to the implementation of the Health and Safety at Work Act 1974 (HSWA).

The Society of Civil Engineering Technicians

Although the Institution of Civil Engineers was founded in 1818 and is the oldest professional institution in the world, the role of technicians was not recognised until 1968, when the Society of Civil Engineering Technicians (SCET) was established.

SCET recognised that technicians undertook work of a 'very important nature' and, because of this, their employers found it necessary for the individuals to undergo formal training (Forester, 1987). They established two formal qualifications: the Technician Engineer, TEng MSCET, and the Technician Fellow, TEng FSCET. Over the next decade this society grew in stature and significance. In the 1980s it looked to merge with the ICE. The full merger came in 1989, bringing over 4000 new members to the ICE (SCET, 1989).

In 1995 the title had evolved to EngTech TMICE before becoming the more familiar EngTech MICE in 2015, which is the title currently in use. The combining of SCET and the ICE has had the benefit of reflecting how technicians are members of the civil engineering community, but perhaps it has also resulted in a partial loss of the independent identity which was achieved through SCET.

This may have given the impression that the qualification of EngTech was subordinate to, or a stepping stone to, status as an engineer as either Incorporated Engineer, IEng, or Chartered Engineer, CEng. While, for some, the opportunity of progressing from a technician to an engineer may be appealing, for many it is not.

In addition, in the late 1990s the industry focus was not on technicians. Educational requirements for IEng were changed and the quantity of people looking to achieve recognition as a technician declined. This, coupled with advances in technology such as CAD, appeared to sound the death knell for the role of technician. But in recent times, like a phoenix rising from the ashes, there has been a renewed focus by the civil engineering industry on technicians and we now have a thriving community of technicians at EngTech MICE and EngTech FICE.

Did you know?

Technicians are defined as:

worker[s] trained with specialist skills, especially in science or engineering.

(Cambridge English Dictionary, 2024)

Revival

This desire to identify and recognise technicians started to build slowly in 2009. It was given a boost by Prime Minister David Cameron when he launched the EngTechNow campaign in June 2013. EngTechNow was led by the Institution of Civil Engineers (ICE), the Institution of Mechanical Engineers (IMechE) and the Institution of Engineering & Technology (IET), in conjunction with the Engineering Council. It was further supported in 2014 by Lord Sainsbury's Gatsby Charitable Foundation (New Civil Engineer, 2014).

In 2011, the Institution of Civil Engineers introduced the Jean Venables Medal. The aim of the award is to promote awareness of the role and achievements of newly qualified engineering technicians. Jean Venables (144th president of the ICE) has significantly contributed to raising the profile of technicians. Her efforts have left a lasting impact, ensuring that the contributions of technicians are acknowledged and celebrated within the engineering community.

This newfound recognition, coupled with a growth in college-based apprenticeship courses such as civil engineer technician and civil engineer senior technician, has fuelled a dramatic enthusiasm for recognising the considerable value technicians bring to the industry. Since 2009 the number of people achieving technician membership of the ICE has increased from an average of 50 per year to 400 per year. Although there is still work to be done, there has clearly been a positive shift towards making technicians more visible and better supported.

> "I'm delighted to see technicians getting their own guide to help them prepare for the technician review. This is a class of membership that did not exist when I wrote *Preparing for the Professional Examinations of the Institution of Civil Engineers* in 1989.
>
> Technicians are an integral part of a project team and being a Technician Member of the ICE demonstrates your competence and observance of professional standards."
>
> **Jean Venables CBE FREng FICE FCGI, 144th President of the ICE (2008/2009)**

The technician community must be respected and valued within a supportive and inclusive environment. The need for the civil engineering industry to grapple with and define the difference and commonality between technicians and engineers is crucial. This will inevitably improve how the industry professionally develops in the future. Professional recognition and the commitment that comes with it will visibly raise the status of technicians and show the vital role they play in the successful delivery of civil engineering projects.

REFERENCES

Cambridge English Dictionary (2024) https://dictionary.cambridge.org/dictionary/english/technician (accessed 05/09/2024).

Forester G (1987) *Building Organisation and Procedures*, 2nd edn. Routledge, London, UK.

New Civil Engineer (2014) https://www.newcivilengineer.com/archive/technicians-campaign-names-new-boss-19-05-2014/ (accessed 05/09/2024).

SCET (Society of Civil Engineering Technicians) (1989) *Minutes of the SCET Amalgamation Committee 1968–1989*, held in archives library at ICE, London, UK.

University of Bristol (2023) https://researchculture.blogs.bristol.ac.uk/2023/07/26/celebrating-technicians-a-look-at-the-past-present-and-future-of-technical-roles-in-higher-education-and-academia/ (accessed 05/09/2024).

Malcolm Peake
ISBN 978-1-83549-943-6
https://doi.org/10.1108/978-1-83549-940-520251002

Chapter 2
Technicians

Who are technicians?

It may seem strange to start the chapter with 'Who are technicians?' in a book that is for technicians. But for some people working in the civil engineering industry, it is difficult to pinpoint just who the technicians are. To define all the roles and job titles of a technician is impossible. A booklet written by Hannah Shewan-Friend *et al.* (2023) is an excellent guide and goes a long way to explain things.

This is not a story book. It is not intended for you to read it cover to cover. If you are eager to get started with your application, then don't delay; jump to chapter 5 and get the information you need so you can put your application together.

This chapter is aimed at giving you some further detail and information to provide you with clarity and confidence, if confidence is needed, that the work you do and the experiences you have are right for someone looking to become qualified as a technician member of the Institution of Civil Engineers. There are some case studies and personal testimonies included at the end of this chapter. They provide some real-world examples to help make the application process more approachable.

Technicians and engineers are two career options that share many similarities. Some people use the terms technician and engineer interchangeably. This creates confusion. There are differences in these roles. In this book we will attempt to capture some of the differences and technical expertise that technicians from across the civil engineering industry undertake.

Engineers and technicians have different responsibilities. Civil engineering projects are completed successfully when engineers and technicians work together side-by-side to identify problems and devise solutions to fix those problems together.

Technician members work in a wide variety of roles within the civil engineering industry. Not everyone is the same; they work at a variety of levels including technical, supervisory, management and in some cases are the directors of their own companies.

Civil engineering technicians contribute in many ways to different infrastructure projects, such as highways, railways, surveying, water resources, buildings, structures and geotechnical. They work in different types of employment, including as consultants, in construction companies, for infrastructure owners and at academic institutions.

Did you know?

Civil engineering technicians work in:

- bridges
- environmental planning / engineering
- airports
- geology
- geotechnical and round engineering
- tunnelling
- offshore engineering
- transportation planning
- regeneration and development
- dams
- reservoirs
- building
- structures
- water supply
- sewerage treatment
- drainage and networks
- railway systems and infrastructure
- river
- coast, marine
- docks, harbours
- highways and traffic engineering
- energy services

(ICE, 2024c)

Technicians undertake many different roles. Some are involved in constructing roads, inspecting rail infrastructure or working on 2D/3D CAD models alongside engineers. Other technicians are independent specialists working in laboratories or provide other trade services, such as specialist skills in material testing or scaffolding design. Technicians may be directly involved in the delivery of civil engineering projects on site as a manager, supervisor or in an appointed role.

Civil engineering technicians work on a wide variety of infrastructure projects. They undertake independent technical work and, at times, provide support to engineers and other construction professionals in the *construction process*. They may work in a range of specialist areas; they can work independently or as part of a team of civil engineering technicians, working alongside or at times reporting to senior civil engineers to coordinate the delivery of components of complex projects.

Did you know?

The construction process could include:

- conception
- design
- construction
- commissioning
- maintenance
- decommissioning
- removal
- management and procurement.

Experienced technicians often hold significant and influential roles, leading not just to the physical delivery of projects but also checking the quality of work done and developing professional abilities in others. If in doubt about your suitability you can always contact one of the membership recruitment team at the ICE and they will be happy to help.

> "From starting as an apprentice to now mentoring our future EngTechs, I have gained a great appreciation for the work that goes into our development. I am glad to be a part of helping grow the future engineering workforce."
>
> **Alex Hudson EngTech MICE**

Academic qualifications

Some technician positions require you to hold a secondary school certificate or the equivalent. Typically, this refers to education provided to GCSE level (General Certificate of Secondary Education). Commonly, people will then go on to develop their academic knowledge through further education (FE) colleges with a BTEC and on to higher education (HE) courses such as level 4 diplomas, HNC, HND or similar qualifications. Increasingly there are opportunities to develop careers using level 3 or level 4 apprenticeships.

Other technician positions require a Bachelor's (BSc or BEng) degree from a university or college. There are other ongoing educational opportunities that technicians can pursue to maintain their expertise and keep up with the latest trends in the industry. For example, this could include level 6 NVQ in Construction Site Management, MSc in Building Information Modelling (BIM) or level 7 NVQ in Construction Senior Management.

Did you know?

To apply for EngTech MICE at the ICE, ordinarily you will need one of:

- National Certificate/Diploma in Civil Engineering (inc. NVQ level 3 in technical or construction subjects)
- HNC/HND in Civil Engineering
- Level 3 or level 4 apprenticeship in Civil Engineering.

Professional qualifications

Becoming professionally qualified as a technician is a worthwhile goal. It allows people to become more visible within their organisation. You become role models for your peers and coworkers. Others will take notice that you have reached a recognised professional level. It demonstrates to prospective clients the level of skills and professionalism that a company has when they are bidding for work.

Depending on the specific roles and responsibilities, technicians can use a blend of creative technical and suitable mathematical skills. The engineering technician qualification is for anyone in the construction industry who carries out technical work competently, safely and independently.

> "Professional technicians are skilled individuals. They are active members of the team. Their input and feedback are essential to delivering efficient, high-quality engineering solutions either during construction activities on site or when building engineering content in the design office."
>
> **Tom Price EngTech FICE ACIArb**

Routes to professional qualification

There are three primary routes to achieve the technician qualification. The first two pathways are the direct application for technician professional review (TPR), which reflects on the work-based development that often takes place after completing academic studies, and the modern apprenticeship through an end point assessment (EPA) which combines an assessment of practical work experience alongside academic knowledge. But you don't need to have been to college or have any academic or vocational qualifications to apply. The third pathway places value on work-based learning where skills are gained through a combination of training, experience and learning on-the-job. This is commonly known as 'experiential learning'. The review is sometimes known as the 'technical technician professional review'. But more commonly it is simply referred to as the technician professional review (TPR).

> "As someone who took a hands-on practical route rather than a formal educational one when leaving school, to be professionally recognised by ICE as EngTech FICE was a highlight of my career thus far and something I'm immensely proud of.
>
> Gaining professional recognition has given me opportunities that I'm convinced would otherwise have not been available to me. It has allowed me to show and promote my particular field (scaffolding) for the engineering discipline that it is."
>
> **Steve Byrom EngTech FICE**

Technician professional review (TPR)

As a part of the review, you will be assessed against seven Attributes. These set out the types of behaviours expected of a professional technician. These are described below. To find out more about the Attributes, turn to chapter 4; putting your submission together, go to chapter 5 and for details about the day of the review, turn to chapter 6.

Apprenticeships

These are a new way for people to gain an academic qualification and professional recognition at the same time. The apprenticeship training ordinarily lasts for 36 months, concluding with an end point assessment (EPA). As an integral part of the EPA, you can also apply for recognition as a civil engineering technician (EngTech MICE). For further details on apprenticeships, go to chapter 7.

Experiential learning

If you do not have a formal qualification but you have built up your knowledge through experience on the job, you can still apply. Your review is extended and the extra time is used to confirm that the experience and knowledge gained through work-based experiences are equivalent to those of an applicant with an approved qualification. There is more on this route in chapter 6.

Did you know?

The seven Attributes are:

- understanding and practical application of engineering
- management and leadership
- commercial ability
- health, safety and welfare
- interpersonal skills and communication
- sustainable development
- professional commitment.

A key part of justifying someone's credibility often lies in a person's qualifications. These cannot be viewed in isolation and other relevant factors should be considered. But holding the status of professional technician and member of the Institution of Civil Engineers enhances your professional credibility. It could open doors to new opportunities and can give you access to resources and events for continuous professional development.

Summary

Many people apply to join the ICE during the initial stages of their career. However, it can be just as beneficial to apply at any time in your career. Many people in the civil engineering community hold a wealth of valuable, hard-won experience and the more of them who get on board with the ICE, the better. The ICE seeks to continue to diversify its membership while holding up the standards of professional behaviours. Key to addressing industry concerns about safety, quality, cost and sustainability will lie in building up the diversity of the membership. By embracing and including technicians, the ICE will reflect a wider range of professionals involved the industry.

> "For me the ICE is the pillar I lean against when I need practical engineering back-up – there is always someone that has been in the same situation and is willing to share their knowledge and wisdom."
>
> **Aaron Passfield EngTech MICE**

Application documents comparison

Table 2.1 summarises the different application documents associated with the four main routes that need to be submitted to achieve EngTech MICE at the Institution of Civil Engineers.

Table 2.1 Summary of application documents (ICE 2024a, 2024b, 2024c)

Stage 1: initial application

	EPA level 3 v1.1 (CET)	EPA level 4 v1.1 (CEST)	TPR	TPR (missing Academic Base)
Application form	✓	✓	✓	✓
Attributes	N/A	N/A	3000 words	3000 words
Portfolio of evidence (KSBs)	✓	✓	N/A	N/A
CPD	✓*	✓*	✓	✓
Sponsors	✓*	✓*	✓	✓

*only applicable if applying for professional recognition as EngTech MICE

Stage 2: technical project brief (TPB)

	EPA level 3 v1.1 (CET)	EPA level 4 v1.1 (CEST)	TPR	TPR (missing Academic Base)
TPB presentation	10 min	10 min	N/A	N/A
TPB report	2500 words	3500 words	N/A	N/A

The presentation and report should be combined and submitted as one pdf.

Assessment/review day comparison

Table 2.2 summarises the similarities, differences and details that affect the review/assessment day associated with the four main routes to achieve EngTech MICE at the Institution of Civil Engineers.

Table 2.2 Summary of details of the four different assessment days

	EPA level 3 v1.1 (CET)	EPA level 4 v1.1 (CEST)		TPR	TPR (missing academic base)
Part 1 (TBP): 30 min	10 min presentation	10 min presentation	45 minutes	5 min (optional)	5 min (optional)
	20 min Q & A (recorded)	20 min Q & A (not recorded)		Up to 45 minute Q & A interview on whole application to confirm you meet all the attributes.	Up to 60 minute Q & A interview on whole application to confirm you meet all the attributes with focus on understanding and practical application of engineering
Part 2 (PoE): 40 min	40 min Q & A on remaining KSBs (not recorded)	40 min Q & A on remaining KSBs (includes greater emphasis on BIM and evaluation of digital modelling techniques) (not recorded)	15 min		Extra time
Result	Distinction / Pass / Fail		Result	Pass / Fail	

Personal testimonies

There is a vast array of roles that civil engineering technicians undertake. It would be impossible to cover or even define all the possible roles and job titles that technicians do. This series of personal testimonies has been brought together to provide some real-world examples and inspiration.

These testimonies come from a diverse group of people working in a range of different roles. They have all taken the time to share an insight into their typical working day, why they applied for the EngTech MICE qualification and what advice they would give to someone who was to apply for a technician professional review (TPR) or end point assessment (EPA).

Sam Ashman EngTech MICE	
Education and qualifications:	BTEC Construction and the Built Environment (Civil Engineering) and HNC in Civil Engineering
Job title:	BIM Manager
Describe a typical working day:	My day-to-day role as a BIM Manager involves a variety of different things. Responsibilities can vary massively.
	My duties include delivering design drawings and models, coordinating with stakeholders on Tony Gee projects, from bid stage to completion, and facilitating visualisations with clash detection when needed.
	I train project teams in various software packages and standards to meet client requirements, conduct audits to ensure compliance with CAD/BIM standards and assist in preparing bid proposals and key project documents (MIDP, TIDP, BEPs).
	I collaborate with multidisciplinary teams, regional BIM representatives and global offices, while upskilling staff in BIM and CAD. I also support staff aiming for EngTech certification and ensure our BIM procedures remain up to date and compliant with BS EN ISO 19650.
Why did you apply for professional registration as EngTech MICE?	It opens many doors in the industry and there are many different career paths you can take. EngTech showcases that I have the capability and knowledge needed to carry out my job and allows me to demonstrate this to my peers.
	I hope to progress to CEng in the future, should there be a suitable route for technicians.
What advice would you give to someone considering professional registration as an EngTech MICE?	Go for it – it shows you are a competent individual and opens many doors and opportunities in the industry. It's also great for networking and meeting other like-minded people.

Rob Ehren EngTech MICE	
Education and qualifications:	Diploma in Construction in Built Environment and HNC in Civil Engineering, Construction Engineering
Job title:	Asset Management Officer
Describe a typical working day:	Each day is different. Ordinarily I organise asset maintenance work and meet with field teams to check on progress.
Why did you apply for professional registration as EngTech MICE?	The apprenticeship helped me expand my knowledge in the field and produce a higher standard of work. Getting recognised by the ICE was a natural next step.
What advice would you give to someone considering professional registration as an EngTech MICE?	Start the ball rolling and see what you think. I think it offers a lot of opportunities to meet people in the industry that you may not meet otherwise. There are also a lot of events that are held that have interesting people speaking.

Rich Tournay EngTech MICE	
Education and qualifications:	NVQ level 2 Plant Operation, Diploma in Construction in Built Environment and HNC in Civil Engineering, Construction Engineering
Job title:	Asset Performance Team Leader
Describe a typical working day:	I am responsible for the maintenance and performance of strategic flood defences from flood defence banks and walls to pumping stations and sluices. Each day is different. Ordinarily I organise asset maintenance activities, complete utility searches and liaise with field teams. I also get involved in incident responses during flood events.
Why did you apply for professional registration as EngTech MICE?	Membership of the ICE is recognition of my professional standard.
What advice would you give to someone considering professional registration as an EngTech MICE?	I guess my advice would be: don't be afraid to do it, don't be worried or overthink it. You're in a room with professionals who want you to achieve and show that you're a professional in what you do. Also there are those around who are willing and wanting to help, so don't be afraid to ask.

Jill Mascarenhas EngTech MICE	
Education and qualifications:	BTEC level 3 Construction and the Built Environment
Job title:	Assistant BIM Technician
Describe a typical working day:	On a typical day, I work with CAD software MicroStation to develop 2D and 3D models for various tasks, including typical details and track drainage. I then create detailed drawings from these models to enable the client to see our design intent and the contractor to build from. My role involves close collaboration with engineers and technicians to create, revise and amend these drawings and models according to their mark-ups. I often like to take on tasks where I can see gaps – a recent example of this is developing a document register for an area of the project I am currently working on. I also spend a lot of time supporting apprentices in their day-to-day work and college experience. A recent example of this is creating and providing resources and packs to aid them in their EPA process.
Why did you apply for professional registration as EngTech MICE?	I wanted to demonstrate my knowledge of industry standards and show my dedication to my role, including continuous development.
What advice would you give to someone considering professional registration as an EngTech MICE?	My best advice is to understand that you will not necessarily know everything. It is actually part of the ICE Attributes to understand and accept your limitations but always strive to deepen your understanding. One way of doing this is to ask any questions that come to your mind, no matter how 'silly' you think they may be. Take time to do training in areas where you feel you're not as confident.

Ingrid Valceschini EngTech MICE	
Education and qualifications:	HNC Civil Engineering More recently I have completed an MSc in Building Information Modelling Management & Integrated Digital Delivery
Job title:	Associate
Describe a typical working day:	The beauty of my work is its variety. Each day brings something new as I manage both national and international projects, primarily focusing on bridge design. My responsibilities include assisting engineers in the design process, creating 3D BIM models and developing drawings for on-site bridge construction. Additionally, I plan the digital strategy for BIM projects, develop BIM documents and support other technicians in understanding and delivering these projects. My daily tasks also involve resource planning for my technician team, training the team and recruiting new apprentices.
Why did you apply for professional registration as EngTech MICE?	I applied for professional registration as EngTech MICE because my company encouraged and supported me. Becoming professionally qualified has significantly enhanced my career prospects and professional development. It led to my promotion to Associate and provided the opportunity to join the MSc in Building Information Modelling Management & Integrated Digital Delivery. My journey continues, as I have recently applied to become an ICE reviewer.
What advice would you give to someone considering professional registration as an EngTech MICE?	Becoming a professionally qualified technician as an EngTech MICE can significantly boost your career, opening up new possibilities for both career advancement and further study. It's a fantastic opportunity to deepen your understanding of engineering and gain recognition for your skills. Think of the registration process as a constructive discussion with professionals rather than something intimidating. Embrace this journey with confidence and enthusiasm, knowing that it will pave the way for exciting new opportunities in your professional life.

Annabel Smith EngTech MICE	
Education and qualifications:	BTEC level 3 Construction and the Built Environment
Job title:	Assistant BIM Designer
Describe a typical working day:	On a day-to-day basis I work primarily on Autodesk Revit creating in-depth structural 3D models of buildings and creating 2D drawings from the 3D models. I usually work on projects from work stage 2, 3 and 4 up until construction record. I work alongside other technicians and work closely with engineers and update drawings and 3D models as per engineer's mark-ups.
Why did you apply for professional registration as EngTech MICE?	I applied for professional registration as EngTech MICE as I wanted to demonstrate that I have the competence and understanding of everything that I had learnt at work and at college.
What advice would you give to someone considering professional registration as an EngTech MICE?	The advice I would give to someone considering applying for EngTech MICE is to ask questions at work. Don't be scared about asking questions and wanting to understand more if something isn't making sense.

Be interested in learning more about what you are working on and how everything comes together. Site visits are really helpful for this. You don't need to know everything but it's great to have a basic understanding of as many things as you can in preparation for the interview. |

Amy Pledge EngTech MICE	
Education and qualifications:	HNC in Construction
Job title:	BIM Designer
Describe a typical working day:	I work alongside the engineers within the marine team to produce their designs on schemes using software such as AutoCAD, Civil 3D, WINFAP and HEC-RAS.
Why did you apply for professional registration as EngTech MICE?	My first job was working as a receptionist. When I changed my role and started working as a trainee technician, I was introduced to the ICE and the professional qualifications. I liked the idea of the EngTech qualification but was not confident in thinking that I would achieve it. I studied at college and spent a few years gaining the experience that I needed before deciding to sit my professional review in 2022.
What advice would you give to someone considering professional registration as an EngTech MICE?	It is ok to feel overwhelmed and scared about the process. Ask for help and advice from your sponsors and the people around you and talk to people who have sat their technician professional review. Ask your sponsors to sit mock interviews with you so that you can practise beforehand to ease the nerves, ensure you are happy with your application and presentation, and to understand the interview process. This is a process that, despite feeling really nervous about, helped me to calm my nerves and ensure that my application was ready for the official professional review.

Ibrahim Kapasi EngTech MICE	
Education and qualifications:	BEng (Hons) Civil Engineering with Site Management
Job title:	Section Engineer
Describe a typical working day:	As a section engineer, no two days are ever the same. My role involves planning, coordinating and leading the safe delivery of major works on site. I am responsible for ensuring that tasks and works are done to the highest quality in line with all safety and technical requirements. I am also the first port of call for any queries the site team may have. I regularly liaise with numerous teams within a construction project, such as health and safety, environmental and commercial. As an engineer, you become the hub of information on the project at hand.
Why did you apply for professional registration as EngTech MICE?	Becoming EngTech was my first stepping stone to becoming a Chartered Engineer. It was also intertwined with my apprenticeship, which made it a no-brainer to complete. The EngTech application provided me with useful skills I could implement throughout my career, so it was never thought to be wasted effort.
What advice would you give to someone considering professional registration as an EngTech MICE?	Don't think of it as an option, just do it! You'll never look back and regret it. It only benefits you. It's a title you carry for life and makes you stand out in a crowd. The process seems difficult, but as you get into it, it becomes easy and achievable.

Sebastian Kreft EngTech MICE	
Education and qualifications:	I studied BTEC level 3 in Civil Engineering, then HNC level 4 and HND level 5, both in Civil Engineering
Job title:	BIM designer
Describe a typical working day:	My day-to-day role mostly consists of delivering multidisciplinary rail schemes throughout the UK. I also dedicate time to provide training to staff as well as research BIM development, where applicable. I primarily work with Microstation and Projectwise; however, I also use other software packages such as AutoCAD.
Why did you apply for professional registration as EngTech MICE?	After I had completed all my studies, I focused on my career development within my role and applied for EngTech MICE. It shows my capabilities and understanding, which helps me gain recognition internally and externally.
What advice would you give to someone considering professional registration as an EngTech MICE?	I encourage people to go for EngTech to show they are a competent individual and strive for further development within their career.

Tom Player EngTech MICE	
Education and qualifications:	HNC in Civil Engineering BEng (Hons) in Civil Engineering
Job title:	Engineering Safety Lead for Data Centres
Describe a typical working day:	I spend most of my time engaging collaboratively with project delivery teams, designers, sub-contractors and clients on engineering in health and safety, and engineering out risk through design changes, product selection or change of sequence in the programme. Once we've found engineered safety initiatives, we then work through them as a team to implement the changes, if required, and then showcase the changes / learning through engineering health and safety case studies.
Why did you apply for professional registration as EngTech MICE?	I applied for EngTech MICE as it was a way of validating my current level of technical ability and allowed me to show colleagues who hadn't met me yet what my level was. I found going for EngTech MICE daunting, I was never particularly academic, and didn't think I had enough experience to go for it. However, if I'd had a book like this one, I would've realised a lot sooner that I was more than capable to go for EngTech MICE.
What advice would you give to someone considering professional registration as an EngTech MICE?	It's a fantastic opportunity to show off your ability to all involved in engineering, the ICE and your work colleagues. Doing EngTech MICE has set me up well in terms of understanding the process of professional reviews, putting me at ease going for IEng or CEng in the future. Use your sponsors or colleagues who have sat a professional review. A mock interview will put your mind at ease for the actual interview.

Hannah Shewan-Friend EngTech FIHE MCIHT MICE	
Education and qualifications:	BTEC National Certificate Civil Engineering (level 3) and a BTEC Higher National Certificate Building Surveying (level 4)
	I am currently studying for a Master's Degree in Building Information Management (BIM)
Job title:	Senior Engineering Technician / Information Manager
Describe a typical working day:	As a senior engineering technician and information manager, my typical day is a blend of technical expertise and data management. I review project updates and address any urgent issues that may come in from contractors, other design team members or our clients. Throughout the day, I split my time between overseeing technical aspects of the project, such as analysing data or reviewing design specifications, and managing the flow of information across various teams. This could involve updating project databases, coordinating with different departments to ensure seamless data sharing, and preparing reports for stakeholders. I also find myself trouble-shooting technical problems, mentoring junior staff, both technician and engineer, and participating in project meetings to discuss progress and challenges. My role requires a keen eye for detail, strong communication skills and the ability to balance hands-on technical work with high-level information management responsibilities. When I am not juggling all of this I am also stepping in and 3D modelling or setting up drawings as well as studying for my Master's.
Why did you apply for professional registration as EngTech MICE?	I put it off for years as I found the whole process quite daunting, if I am being honest. I am neurodiverse so I found being judged and having to prove my worth really difficult. I wanted to sit my professional qualifications because I wanted to be able to demonstrate that I am professionally competent at what I do. It gave me an edge when I was running my own business full time as it showed my clients that I was serious about my company and the work that I undertook and it gave my clients confidence that they were working with a competent professional. Although I found the process hard, I did it, I didn't let my neurodiversity stop me – in fact I have learnt to realise my neurodiversity helps me see things that others miss. My EngTech has enabled me to demonstrate on every project that I have worked on that I am a highly skilled and qualified professional.
What advice would you give to someone considering professional registration as an EngTech MICE?	Be brave, have confidence in your abilities, don't be afraid to reach out to your colleagues and ask for help and advice. Take every learning opportunity thrown at you because it really is a rewarding experience to achieve this goal. Sit as many mock reviews as you can to build your confidence and ask lots of questions. If, like me, you are neurodiverse, don't be afraid to say so because reasonable adjustments can be made to support you, your style of learning and to help build your confidence.

Mike Clark EngTech MICE	
Education and qualifications:	BTEC HNC in Civil Engineering and ISO 19650 Information Management Certification
Job title:	BIM Coordination Manager
Describe a typical working day:	My responsibilities are highly varied from one day to the next. I have a passion for BIM. I'm often involved in supporting the implementation of BIM on projects and helping contractor clients understand the BIM requirements imposed on them. I actively promote the benefits of BIM and how they can improve the project quality. I often work with our clients to agree the appropriate level of information needed and then implement a proper workflow process as well as manage the common data environment. I regularly provide input into tenders to showcase our BIM capability and review project documentation to ensure BIM activities are adequately accounted for. I deliver training on compliance and develop integration with existing processes. This includes guiding design teams in their digital deliverables.
Why did you apply for professional registration as EngTech MICE?	I applied for EngTech MICE so that my technical capability could be assessed and validated by my peers. I also sought recognition for the value I bring to this industry. I became an EngTech reviewer shortly after to help maintain a high standard of EngTech members and enable me to support others through the process.
What advice would you give to someone considering professional registration as an EngTech MICE?	Take it seriously. Do not look at it as an easy pass but as something that gives you the opportunity to showcase your technical competency and value to the industry. It demonstrates your commitment and will ensure you have developed (or been given the opportunity to develop) wider skills beyond just the technical.

REFERENCES

ICE (Institution of Civil Engineers) (2024a) *Civil Engineering Technician (CET) Apprenticeship Version 1.1, Version 5, Revision 6*. ICE, London, UK.

ICE (2024b) *Level 4 Civil Engineering Senior Technician (Version 1.1) End Point Assessment Application Form, Version 1, Revision 3*. ICE, London, UK.

ICE (2024c) *Technician Professional Review Application, Version 3, Revision 6*. ICE, London, UK.

Shewan-Friend H, Brownbridge T and Price T (2023) *Calling All Technicians*, 2nd edn. ICE, Kent & East Sussex Branch, Kent, UK.

Malcolm Peake
ISBN 978-1-83549-943-6
https://doi.org/10.1108/978-1-83549-940-520251003

Chapter 3
Mentors and sponsors

Introduction

There is a general lack of familiarity in the industry in knowing just what it takes to apply for and pass a technician professional review (TPR). For example, the approach to the review is different from that usually undertaken by people aspiring to be recognised as an Incorporated Civil Engineer or a Chartered Civil Engineer.

There are many people involved in the training, development and mentoring of engineers. This creates a natural support network with each successful generation mentoring the next. This culture is currently not as well defined for technicians. This is one reason that a chapter on the subject of mentors has been included in this book.

Mentors often serve as role models, offering valuable feedback and fostering a positive learning environment. This role can be rewarding. Through regular interactions, a mentor helps mentees build confidence, expand their networks and unlock their full potential. Mentoring allows people to pass on knowledge and experience, contributing to the profession and leaving a lasting impact.

For further information on how engineers can be successful, it is recommended to read *Successful Professional Reviews for Civil Engineers* by Patrick Waterhouse (Waterhouse, 2022). This was first written by H Macdonald Steels and published in 1997. It is generally regarded as a comprehensive and indispensable practical guide. While there is a short reference in the book to the TPR and there are some similarities between all the reviews, it is also important to remember the differences.

This book is written with a focus on the TPR. There are significant links between the TPR and the end point assessments (EPA) at level 3 and level 4. If you are an apprentice (or mentor of an apprentice) and want to know more about the EPA, there will be some benefit in reading the whole book. Chapter 7 is where the real detail can be found.

> "The development of technicians is something I have always supported. Acting as a mentor has been inspirational as these professionals realise their potential, often at the start of their professional journey.
>
> To see their skills develop and be benchmarked at a professional review is an immense pleasure and reassuring to see the future of our profession in safe hands."
>
> **Eur Ing Mike Rogers CGeol FGS Eur Geol CEng FICE CEnv FCICES**

As the number of applicants for the TPR and EPA continues to grow, there is an increasing need for mature, capable, competent technicians to take up the role of reviewer or end point assessor. By embracing and including technicians in these roles, the ICE will better reflect the diverse range of professionals involved in the industry who themselves define what makes a technician and therefore set the standards of professional behaviour expected and reflect the standards of the ICE.

This book has been developed to bring clarity to the application process and enable a more diverse group of people to be able to achieve professional recognition as technicians at the ICE. This choice is often made by individuals out of personal commitment. This does not mean the qualification should be sought alone. This chapter is written to encourage applicants to seek out support and appreciate the type of support that will be beneficial to them.

Too many people take one look at the application and react with an overwhelming feeling of dread, with only a vague idea about how to tackle it. This can be the first mistake. When tackled alone, the process can be intimidating. The key to success is identifying the underlying problem and taking steps toward finding a solution. The best way to do this is to not go it alone but to work with a mentor (or two) who can provide suitable guidance and support.

Mentoring requires a significant investment of time and energy. A mentor can provide you with unbiased advice, using their relevant knowledge and experience. Their insights will help you to better understand whether to pursue the idea in the first place or walk away. If you do choose to pursue it and apply, they can help you appreciate what steps you will need to take to achieve success.

Terminology
This book is primarily aimed at those people aspiring to join the ICE as a technician member. In this chapter the terms mentor and sponsor have been used interchangeably. This is not to confuse readers; it is to reflect the duality of the role people often take on. Mentors are people who support trainees; sponsors are responsible for scrutinising a candidate's application and declaring they are happy it contains all the Attributes required by the review.

Mentors
Civil engineering is about working with people to solve challenges. Therefore, the key to solving this challenge (preparing for a successful professional technician review) will be a good mentor. The primary aim of the mentor, in relation to this book, is to support you to achieve professional recognition at the ICE. An ideal person to take on this role would be someone who has sat the review themselves, although this is not essential.

> "My inspiration for becoming professionally qualified was my work with college apprentices. As one of their mentors, I wanted to inspire them by setting an example. By being a professional I could mentor them and sponsor them in reaching EngTech MICE."
>
> **Sebastian Kreft EngTech MICE**

In general, mentors need to be good listeners, patient and able to provide constructive feedback. They must also be adaptable, as your needs and goals will vary widely from others. Your mentor will need to be familiar with the latest ICE guidance to ensure their advice is accurate and up to date. It is for this reason that a whole chapter has been devoted to the topic of mentors. Your mentor will be one of the two sponsors for the review.

Your sponsor must know you professionally and know about the work you do and the standard you are capable of working at. The details behind putting the submission together are covered in greater depth in chapter 5 of this book. The sponsors should also understand the Attributes and appreciate how you have achieved them and applied them in your role(s). In reality, they can only really do this if they have a personal and professional knowledge of you.

Given the lack of formal training schemes for technicians there is a huge amount of work to be done once someone believes they could apply for the TPR. Therefore, the sponsor often naturally takes on the role of mentor as they look to support someone preparing an application. To aid both you as the applicant and your mentor/sponsor, some ideas and advice have been collected in this chapter to help develop a meaningful focus when looking to prepare for a TPR.

Having a good mentor can make a significant difference to your preparation for the review and could help ensure you are adequately prepared. A good mentor will help to ensure you have a well-written application and help you to prepare to demonstrate the required skills. This is a significant investment in terms of time, but it should improve your ability to write about your capabilities and discuss your experiences on the review day, which means all the effort will have been worth it.

Some mentors may look to establish a targeted approach to your preparations. This could include regular fortnightly catch-ups (online or in person). Each of these meetings could look at what you want to achieve and how realistic your goals are. You could work with your sponsors to break down the goals you have and look to find possible solutions to each one to help you move forwards. This type of mentoring is not always possible, but where clear goals can be set then this model can work.

Just meeting up regularly and chatting is not likely to produce any real benefits. Do not be passive – you should also drive the mentoring relationship. After all, you will be the one who has the most to gain from achieving professional recognition.

Finally, a mentor should be someone who has a reputation for instilling confidence in people. This type of mentor can help you establish a few basic rules. They can help you establish where you currently stand, where you want to be and how you are going to move towards achieving your goal and pass the TPR.

> "I was very relieved to pass and felt very proud of my achievement. I think it validates your position in the industry and sets you apart from others."
>
> **Rob Ehren EngTech MICE**

Guide for mentor meetings

Identify aspirations

Right from the first meeting it is important to define what you hope to get out of the relationship and what the mentor is able to do. It is good practice to identify any short-term goals, such as the application date or the month you intend to sit your professional review. Initially you may need a little guidance from your mentor to ensure your goals are reasonable. Once you have a goal you can start to develop a plan. Then you can start to work on achieving your goal. As you progress, be aware: your target may change but there should always be a target to aim for or you may end up directionless.

Set deadlines

By setting targets and actions to hit specific deadlines you will give a sense of urgency to the process. This will help you to focus your time and help you finish the application in a timely manner. However, a large challenge like the application can often seem intimidating, which can easily lead to procrastination. By breaking down the main goal into several smaller steps you can make the workload feel less daunting and therefore more manageable. By setting priorities and breaking the bigger project into smaller tasks you will be able to make a plan to get through the application process.

Set aside time to meet

You will find it far easier to get this done if you purposely set aside a time and location to meet. Having a regular dedicated opportunity to discuss progress, gain feedback and explore challenges will be far more useful than trying to squeeze in a quick chat when the occasion grabs you. These meetings could be a mixture of face-to-face or online. It is the rhythm and regularity that help progress. You should have something to discuss, have an agenda or a question you need answering. You will find this to be very effective and motivating.

Measure progress

Once you have identified your targets and agreed them you should monitor and manage your progress against them. You should celebrate the little wins and if you notice you are slipping behind you can choose how to react. You may choose to apply yourself, work hard and increase your focus on the application or you may reflect on your programme and adjust the ambitions. By measuring your progress, you will be able to acknowledge your progress and ultimately you will achieve your goal in a suitably timely manner.

Sponsors

When applying for professional recognition through either the direct technician professional review (TPR) route or in combination with your end point assessment (EPA) you will need to choose two people to sponsor your application (ICE, 2023, 2024a, 2024b). Sponsors must meet certain requirements, so it's important that you read up on who is eligible to be a sponsor and what they are required to do. They must both be professionally recognised and registered at either EngTech, IEng or CEng grades of membership by a professional engineering institution (for the full range of eligible organisations see ICE, 2024c).

The overriding role of the sponsors is to confirm your suitability for membership. This sometimes involves the difficult decision not to act as a sponsor, which has the potential to be highly demotivating. It could be because you need more experience or simply that they aren't familiar enough with your experience. If you find yourself in this situation then you should ask for a straightforward explanation so you can understand what has led to the decision. This means you can reflect on the advice, fix any issues and adjust your plan for TPR. Clearly there will be short-term challenges, but in the long term this will be helpful and will make you better at understanding your strengths and abilities.

The sponsors must submit their statements of support one week before you submit your application. The implication of this is that before your sponsors can complete their forms, you must have completed (pending minor changes) your submission documents. Bearing this in mind and taking into consideration all the logistics involved in asking them if they are prepared to sponsor you, sending them the forms and checking your final submission, you should consider approaching them no later than four to six weeks before your target date for application. Ideally, if you have worked closely together, they will be aware of your plans.

The current pass rate for TPR at the ICE is very high. This is not because it is easy. It is because of the robust checks that are in place prior to someone making an application. Most sponsors take their responsibility very seriously. One reason for failure in the review can be a lack of experience. This is considered unusual. It is rare for someone with insufficient experience to find a pair of sponsors to support their application.

One of the most difficult tasks of the sponsors is advising an applicant on when to sit their review. The ability to give and receive unwelcome advice is an indication of how good a relationship you have with your sponsor. You should seek out sponsors who will give you honest guidance. Their role is to confirm your suitability for membership.

Did you know?

To protect the values and standards of the ICE, sponsors must be professionally qualified and registered at the EngTech, IEng or CEng grades of membership with the engineering council. All candidates must nominate one of their sponsors as the lead sponsor. The lead sponsor must be recognised as a MICE/FICE.

Lead sponsor

The lead sponsor plays a crucial role in supporting candidates throughout the professional review process. They must be a member of the ICE. The Institution expects the lead to be familiar with the current requirements. The lead sponsor must complete all sections of the statement of support and provide a statement to describe you, your skills and your abilities.

They should describe from their own personal knowledge why they consider you possess the requisite abilities and characteristics, and that you are a 'fit and proper person' suitable for admission to membership of the Institution (ICE, 2024c).

Therefore, a good sponsor is not someone who simply signs off your submission without even seeing the documents. This will not help you be at your best on the day. They should understand the process you are undertaking. They should be willing to set aside time and effort to support you.

Your lead sponsor must be satisfied that you have a realistic chance of proving to the reviewers that you have become a professional technician. So, they should first satisfy themselves that your submission demonstrates all the Attributes. This, combined with your own determination to succeed, should enable you to achieve your ambitions in a timely manner. There is more information on the details of the role in chapter 5 which you should read as you put your submission together.

Second sponsor

The second sponsor has little more to do than simply confirm, by provision of their signature, that you possess the skills and characteristics comparable to an EngTech MICE and you are a 'fit and proper person' (ICE, 2024c). If they wish to, and I would encourage it, they can also provide a short statement of up to 500 words, but this is not mandatory.

Summary

Helping someone grow and succeed can be incredibly rewarding. Good mentoring can be beneficial to everyone. It advances a mentor's own leadership, communication and interpersonal skills. It's good ongoing professional development. It's a meaningful way to create a lasting legacy. Once you have been successfully recognised by the ICE you may find it appropriate to give assistance to others who wish to do the same.

REFERENCES

ICE (2023) *Technician Professional Review Guidance, Version 3, Revision 7*. ICE, London, UK.

ICE (2024a) *Civil Engineering Technician Apprenticeship Version 1.1, Version 5, Revision 6*. ICE, London, UK.

ICE (2024b) *Civil Engineering Senior Technician Apprenticeship Version 1.1, Version 1, Revision 3*. ICE, London, UK.

ICE (2024c) *Sponsor's Statement of Support, Version 6, Revision 5*. ICE, London, UK.

Waterhouse P (2022) *Successful Professional Reviews for Civil Engineers*, 5th edn. ICE, London, UK.

Malcolm Peake
ISBN 978-1-83549-943-6
https://doi.org/10.1108/978-1-83549-940-520251004
Copyright © 2025 by Malcolm Peake. Published under exclusive licence by Emerald Publishing Limited

Chapter 4
The Attributes

Introduction

The Attributes are based on the UK-SPEC guidelines provided by the Engineering Council (EC, 2020). The UK-SPEC describes the requirements that must be met for registration. Technicians work in a wide variety of roles within the civil engineering industry. Not everyone is the same.

This chapter has been written to provide some examples, case studies and ideas on what the Attributes mean to different people who work in the civil engineering industry. People working as technicians undertake different activities across varied roles, such as site supervisors or inspectors of infrastructure. Some people create drawings or models. Others work in universities or material testing laboratories. As a result, this is by far the largest of the chapters in this book.

The first time you look at the Attributes of an EngTech MICE in the ICE's *Technician Professional Review Guidance* document (ICE, 2023b), you may find them quite overwhelming. With practice, the Attributes will become more recognisable and will get easier to understand. The application form splits the skills used by professionals into seven Attributes. You will need to provide information to demonstrate how you use all of them in your ordinary working life.

To help you prepare your application, this chapter starts off with a couple of case studies. Then it has been split into the seven Attributes. Within each section there is an interpretation of what the Attribute could mean and there are further examples related to the main four employment types:

- contracting and construction
- consultancy and design
- academic research
- infrastructure owner/client.

> "EngTech MICE impacts on the individual by raising their self-esteem, reinforcing their belief in what they can do and the contribution they can make to projects by openly demonstrating their professionalism using the post-nominals. It also impacts on their employer by demonstrating the professionalism of the company in the quality and standard of its employees."
>
> **Susan Hawkes EngTech FICE FWES**
>
> **Reviewer and End Point Assessor**

At the end of the chapter a range of case studies have been brought together. It is recommended you reflect on the requirements, interpretations, sample responses and case studies. You will need to work with your mentors to adapt them and apply them to your own experiences.

Apprentices

If you are on a level 3 or level 4 apprenticeship the term 'Attribute' isn't used. You will be more familiar with the term 'learning outcomes'. Similarly, the assessment is called an 'end point assessment' or EPA rather than a technician professional review (TPR). This chapter will focus on the Attributes and the TPR. You can find out more about how they link to the apprenticeship standards in chapter 7.

Expectations

Do not fall into the trap of wondering what the ICE expects from your TPR. The straightforward answer is that it depends on you. There are many ways to demonstrate the Attributes. The more you read and think about them, the more familiar you will get with them and the more you will be able to express yourself.

> "Achieving EngTech status is a significant milestone in a technician's career. It acknowledges their value within the industry and showcases their comprehensive knowledge and skills as a well-rounded engineering technician, validated by their peers."
>
> **Mike Clark EngTech MICE**
>
> **Reviewer and End Point Assessor**

The reviewers will be interested in understanding how *you* apply *your* experience to solve real-world problems. They will also want to know about wider skills like how you organise your day, work safely and work alongside other people. In your application you will be giving examples of what work you can do. This will be the focus for discussions in the interview on the day of your TPR. It is important to make sure you focus on what *you* did, not what *we* did. This makes it clear what your role was and the reviewers know exactly what part you played.

You could think of your application as a portfolio of your recent work. Do not be tempted to describe everything you can and have ever done. Instead, you should focus on your recent experiences. Therefore, pick one or two pieces of work you have recently been doing and use them throughout the whole application. It is important to include examples of your work in your application. This will help you show the reviewers what you do and what records you need to keep. You can expand further on problems you encountered and tasks you have undertaken in the TPR interview.

You can write up to 3000 words. You must target *every* Attribute to prove you have sufficient experience. You do not have to spread the words equally. You should work closely with your sponsors when preparing the application to help you do this.

Attribute 1 Understanding and practical application of engineering

Overview

This is a key Attribute in demonstrating your understanding of engineering and how you apply this to your role. This is at the core of the work we do in our industry and is how you can show your practical knowledge and experience as an engineering technician and how this contributes to the projects you work on.

As an engineering technician, you will be the link between theory and practice, translating engineering designs into reality. This will be your opportunity to provide an example or two of what you can do and how you do it, highlighting the importance of your work and how you use appropriate techniques, procedures and methods to ensure it is done correctly, in a detailed and precise way.

TPR requirements

(a)   Use appropriate scientific, technical or engineering principles

(b)   Review and select appropriate techniques, procedures and methods to undertake tasks

(c)   Identify problems and apply appropriate methods to identify causes and achieve satisfactory solutions

This Attribute has been allocated a number and letters have been added in parentheses to assist with the explanation. These do not feature in the ICE publications.

Interpretation

You will need to show how you apply your understanding of engineering principles to your work, taking what you were taught at college, or have learned through experience, and using it to improve what you do. These principles could be in structures/mechanics, fluids/hydraulics, soils/geotechnical and/or materials. Think about how you have then improved on this experience, honed your skills and how it has helped you to become more proficient in your role.

As a contractor, can you read and interpret drawings? Are you responsible for setting out or surveying a site? Do you follow manufacturers' instructions for the installation of fixings or use of materials? If you are a consultant, think about the work you do assisting the engineers to develop technical solutions. Do you know how to model different materials? Can you produce standard details following design guides? Do you understand how to import different formats to your model and correctly set up coordinate systems? As an academic, are you able to follow a testing procedure accurately and record the results correctly? What common problems occur if a test is not prepared correctly? And as a client, what type of inspections do you undertake of assets? What instrumentation is needed? What procedures and testing do you need to follow to check quality? How do you manage and maintain different assets or property?

You should give an example of the work you do. You should think about your role and the type of engineering you get involved in doing. You are looking to demonstrate that you know what to do and where to find information so you can solve problems. Explain your role and your responsibilities. You should look to use examples that are part of your typical working day. You should highlight some of the work you do and show how you use your technical knowledge to solve everyday problems.

If you do not have approved academic qualifications, then you must place a greater focus on the 'Understanding and practical application of engineering' Attribute when describing your abilities. You will need to emphasise in greater detail how you are able to contribute to solving engineering problems. This is so the reviewers can appreciate the engineering knowledge you have developed through learning 'on the job'. As mentioned previously, you do not have to spread the 3000 words evenly across all Attributes. Therefore, you could choose to allocate more words to Attribute 1 so you can provide more details.

Ideally you will be able to expand on the examples and demonstrate how you apply your understanding of engineering when you select tools, materials and equipment to achieve a satisfactory answer for some of the engineering problems you are responsible for solving.

Sample responses

Contracting and construction: lift supervisor – crane lift

I work closely with the appointed person for a crane lift. I carry out pre-lift checks to make sure the platform has been built according to the specifications and the permit to load is in place. I work with the crane operators to make sure outriggers are placed correctly. I check the condition of chains and shackles that have been delivered and arrange for them to be replaced if there are defects.

I brief everyone involved in the lifting operation on the contents of the lift plan to ensure everyone understands the risks associated with the lift, the control measures in place and their role in the lifting operation. I also monitor the weather and pause the lift if, for example, the wind speeds are too high.

Consultancy and design: detailed design drawings

I work closely with the structural engineer. I am responsible for producing the final construction drawings for the building. I produce an accurate model and detailed drawing, following relevant CAD standards. I make sure they are geometrically and spatially correct. I have an appreciation of what I am drawing/modelling and how my contribution fits into the wider project. My drawings include notes on the grade of steel, concrete and the mix required so the contractor can order the right concrete.

Academic research: laboratory technician – soil testing

I prepare soil samples for testing. This includes trimming the sample to the correct size and ensuring it is fully saturated. I also calibrate the pressure transducers and set up the triaxial cell to ensure there is a good seal. During the testing I apply the load at a controlled rate and monitor the corresponding stress and strain. I collect the data and process it to determine key soil parameters, such as shear strength and cohesion.

I carry out routine checks to ensure the machine is working correctly. By carrying out annual calibration checks I can make sure there is a traceable history of the triaxial machine producing reliable results that are correct to industry standards.

Infrastructure owner/client: tunnel inspector – asset management
I am responsible for carrying out inspection work along track tunnels. I am responsible for completing the survey record, checking for any signs of wear, damage or potential hazards. This includes looking for cracks, water ingress and any other structural issues that could compromise safety. If I see anything that requires immediate attention it is my responsibility to highlight this to my team leader. My work helps in the planning and maintenance of repair work.

Summary

Being technically capable does not necessarily mean you are carrying out analysis or using equations. This is more usually associated with consultant engineers.

In this chapter an attempt has been made to capture a sample of some of the roles and technical expertise that different technicians from across the civil engineering industry undertake. The examples provided try to illustrate some of the duties that technically capable professionals could be undertaking as part of their ordinary working day. By presenting them here it is hoped you will be able to draw inspiration and find ideas related to your own work.

Attribute 2 Management and leadership

Overview

The words 'management' and 'leadership' may feel like they are associated with directors, project managers or individuals in similar leadership roles. However, technicians often overlook the routine, everyday things they need to organise and manage or accept responsibility for in order to do their work. The TPR is for technicians to prove they can do their job well and follow best practices.

It is in the attention to detail and the contributions you make as an individual where the greatest impact can be made to the success of a particular piece of work or project. If each person in a team takes responsibility for the work they do, it will reduce error, improve quality and get work done on time and to budget!

Therefore, it will benefit you to think not about what the people in management and leadership may do, but what your responsibilities are. What do you have to plan and organise? What processes and quality procedures do you have to follow? How do you make sure your work is delivered accurately and on time?

Interpretation

As a technician you will often be involved in specific elements of work. Responsibilities vary with the scope of the project and size of the employer. Your role may be highly specialised. Different technicians may contribute at different times in different ways to elements of a project.

TPR requirements

(a)   Identify tasks and organise resources to complete them effectively

(b)   Work reliably and accept responsibility for their work or the work of others

(c)   Complete tasks with due consideration for quality

This Attribute has been allocated a number and letters have been added in parentheses to assist with the explanation. These do not feature in the ICE publications.

For example, as a contractor, you may be reading drawings so you can construct various items. As a consultant, you may be managing your time so you can create drawings and models. As a lab technician, you may be planning labs or managing the annual maintenance of equipment. As a client or infrastructure owner, you may be using records to organise and plan inspections or prepare maintenance schedules.

Attribute 2(a) refers to identifying tasks and organising the resources to complete them. The types of tasks you are likely to be responsible for and the resources needed will vary depending on the work you do and your role. It is quite likely that you will be responsible for planning and managing your own time, perhaps looking about one week (maybe three) ahead to help you and others appreciate your workload.

For this Attribute, consider explaining the information you rely on to plan your work or maintain records of completed tasks. How do you know what you need to do each week? How are you briefed? Are there weekly or daily meetings to discuss progress? When you gather information to satisfy this Attribute, consider what evidence you could use to demonstrate how work is planned and managed in your organisation.

Do you attend weekly meetings or daily briefings? Do you have access to weekly programmes? Perhaps a screenshot of a resource chart or task list could help illustrate how you plan your time or manage your own workload. Construction personnel often have a programme of activities to carry out or daily diary sheets to record to keep track of site progress and document everyday.

> "Being recognised as an EngTech demonstrates I have met the rigorous standards set by the Engineering Council. By maintaining my registration, I show how I am committed to maintaining those standards and helping others to develop too. I train both technicians and engineers.
>
> EngTech registration is not just about adding letters after your name; it's about standing out in a competitive field and showing that you are committed to excellence in engineering."
>
> **Hannah Shewan-Friend EngTech FIHE MCIHT MICE**

You will need to have evidence of your organisational skills for Attibute 2(b) and be able to work under your own initiative and as part of a team for Attribute 2(c). You will need to be able to follow procedures. You will also need to be able to make decisions so you can complete tasks on time.

By the time you apply for a TPR you should be experienced enough to be able to work independently, contribute to discussions and be able to check your own work. Think about how you take responsibility for your work. How do you show due care and attention? What actions do you need to take? What does it mean to do 'good quality work'? What evidence do you have that you follow your organisation's quality control processes or procedures? Who countersigns your work or provides witnesses to verify that your work meets appropriate quality standards?

What is your role in making sure your work is accurate and reliable? How do you record the quality of work done? Do you raise nonconformance reports and implement inspection and test plans? How do you collect and file records of the work you do?

BIM is increasingly being used by many organisations to manage information and create a single source of truth for various stakeholders. This is especially true on larger, multidisciplined projects. There are various software packages that have been developed to improve how we share data, access a common data environment and keep all records in one reliable location. How does this affect you?

Sample responses

Contracting and construction: setting out surveyor/site engineer

As the site engineer, I am part of the site management team. I am responsible for establishing the level and survey controls for the site. I set out the works as per the contract drawings. I am responsible for ensuring the plans, drawings and quantities are accurate. I am experienced in the use of total stations. I carry out my own checks, but I also make sure the equipment has a calibration certificate. I am responsible for the quality of my work.

I produce a two-week look-ahead in conjunction with the site agent and I make sure my work is done in a timely manner to suit site progress. I keep accurate site diaries to record work done.

Consultancy and design: draftsperson/CAD technician

As part of my role, I am responsible for organising my own working day. On a regular basis I am asked to estimate the time it will take to complete various tasks, such as drawing production or model integration. I need to keep project managers up to date on what work I am doing so they can plan ahead and resource work across the office. I carry out self-checks on my work and follow the company procedures for check prints before information is issued to clients.

> **Academic research: laboratory technician – concrete cube tests**
>
> I liaise with the course tutor at the beginning of each term to timetable planned experiments and agree a schedule of maintenance for various items of equipment. I keep track of lab supplies and order new materials as needed. My role includes the calibration of machines, such as those used for cube testing. This is important as regular calibration is essential to make sure the machine operates safely and provides accurate, reliable results.
>
> Before each experiment I am responsible for ensuring the event is planned. For example, I check that we have all the necessary materials, equipment and PPE to make sure the experiments can be conducted in the time available.

> **Infrastructure owner/client: asset manager – highways**
>
> Each week I sit down with my line manager to go through our task list. Together we consult inspection reports so we can plan work and target resources. This helps identify what work is ongoing and what needs to be done. I am responsible for identifying the scope of the work for new activities, determining any specific tasks and timelines.
>
> Once an item of work is agreed I then monitor the progress of the work and address any issues promptly to minimise any delays. I carry out regular site visits to check on quality. I document any work done and record any changes to the scope for future reference.

Summary

To sum up this Attribute, think about what it is you are responsible for doing, how you plan your time and how you work with other people. Think about what people rely on you to do and what is ordinarily expected of you after you have built up a suitable level of experience in your role. Reflect on what motivates you and the team you work in to complete various tasks. Finally, consider what processes and procedures you are expected to follow when making sure you complete your work with consideration for quality.

Attribute 3 Commercial ability

Overview

Often when asked about commercial responsibility, many technicians would think they have very little commercial involvement as they typically don't write the contracts and they are not responsible for administrating them. You may think of the project director or project manager as the people who are responsible for the commercial performance of project.

While this may be true on one level, as a technician you will be contributing to the commercial success of a project. Being involved in the delivery of projects, you should know how to organise yourself, identify the resources you need and use them efficiently. This Attribute does *not* require you to have a full and detailed knowledge of different contracts and commercial frameworks which are used in the industry.

It requires you to have the experience and ability to be able to identify what needs to be done, work with others, organise yourself and do what you need to do. The reviewers will want to understand your awareness of your roles and responsibilities. They will want you to demonstrate that you have the ability to organise yourself effectively and use various resources to complete your work.

This could include knowing about the cost of materials or equipment. Depending on your position, this may involve managing a team or a budget. You could provide some evidence of the commercial importance of your role by including a copy of a daily diary or timesheets.

TPR requirements

(a)   Identify, organise and use resources with consideration of cost

This Attribute has been allocated a number and letters have been added in parentheses to assist with the explanation. These do not feature in the ICE publications.

Interpretation

The most common cause of disputes is a failure to properly do the agreed work. For example, as a contractor you will be obliged to keep a diary. By creating and maintaining a detailed and accurate diary you will have recorded exactly what was done, by whom and when. This is especially important on busy sites where there are many different elements of a project being done at any one time. Good diaries can help give a clear and unambiguous record of the work that took place on any given day. What type of records do you keep?

It is essential for construction firms to safeguard themselves against substandard work, while also ensuring that they are paid for the work they have completed. This balance between quality and payment is crucial for the success of construction projects. What is your responsibility? What records do you keep? A worthwhile daily diary is one that includes as much detail as possible. Images can be invaluable. They record the work done. A diary sheet can also record the location of the work, the weather at the time the work was (or was not) done. What do you do?

If you are working as a consultant, you will know that it is vital to keep clear records of billable hours detailing time spent working on projects. By keeping meticulous timesheets, you allow the company to accurately bill clients for the time you spend working on various projects. You should be aware of how you identify and organise your own time. Then on a regular basis you will complete a record of the work done in your timesheets.

You should be aware of the budget linked to the work you are doing. As a consultant, this is often referred to in terms of time, such as 'you have two weeks to deliver the work and don't spend more than five days working on it'. This would imply there is a budget to cover five days' work. It will be your responsibility to identify whether you understand the work you are given and whether you have the information you need and the time necessary to complete it.

It is also important if you are a project-based technician to understand how the project is structured. For example, more detail, accuracy or urgency may be associated with a 'construction' issue compared to work associated with an earlier stage in a project lifecycle like the 'feasibility' or 'concept' stages. With increased use of building information modelling (BIM) this has become an important factor in the delivery of projects.

As an academic, it is important to understand the cost of running the teaching laboratories. You may be involved in the maintenance contracts to ensure that the equipment operates and that you can keep the lab running smoothly. Are you responsible for purchasing supplies?

Are you responsible for maintaining an inventory of stock? Do you have to track and manage expenses? You are likely to need to coordinate and manage experiments and schedule the use of lab equipment to make the most use of the time available. How do you do this? What other commercial challenges do you face?

As a client, you will be involved in the prioritising, planning and budgeting of maintenance work. How are you involved in the decision-making process for planned maintenance? Do you carry out the initial investigation work or do you help define the difference between urgent and planned maintenance work? How is this managed? What other things can affect the decision-making process?

What are your responsibilities? Do you help define the budgets for individual projects, or do you take responsibility for monitoring them to check they are delivered on time? What do you do if changes occur or if additional costs arise? How is that process recorded? Do you need to produce a daily diary sheet to record what work was done, where it was done and who did it? Where are these records kept and why do you need them?

The reviewers want to know whether you are aware of your roles and responsibilities in relation to the work you do. At the very least they want to know whether you can organise yourself and identify and use the resources you need to complete your workload. Depending on your role this may extend to managing a team and could include managing budgets and timescales to achieve project needs. This Attribute also includes the measurement and recording of work done in daily diaries or timesheets to track progress.

Sample responses

Contracting and construction: steel erector
I need to look at the drawings and work out how the steel package will be erected. This includes making sure we have enough people in our team to make sure the project will progress smoothly. I sometimes need to coordinate with other teams in the area to make sure we do not work on top of each other and cause unnecessary delays.
I check deliveries to make sure the right materials are delivered. I also need to make sure the setting out engineer has provided us with all the correct information. Finally, I check the work is done well and to the right quality because any errors can be very costly. It's better to build it right first time. I record most of this information in my site diaries.

Consultancy and design: draftsperson CAD/BIM designer
I work alongside two different design engineers on different projects. It is my responsibility to keep them up to date on a weekly basis on how much work I have to do and when it needs to be done.
I am responsible for making sure I record my time against the correct project numbers. Sometimes there is additional work and when this happens, I need to make sure I record my time separately. This is important as it has an impact on the bills sent to the clients.
I know my hourly charge-out rate includes overheads, profit margin and other costs.

Academic research: laboratory technician
I work with the laboratory supervisor to discuss how to spend the planned budget for the labs. I plan the maintenance and calibration of testing equipment. I make sure to minimise the impact on teaching and research projects.
I am responsible for keeping all concrete cube moulds clean and in good working order. In addition, I am also responsible for manging the PPE store and making sure we have appropriate and sufficient PPE to protect students and researchers when they are working in the labs.

Infrastructure owner/client: tunnel inspector – rail
Each week I am provided with a planned schedule of assets to inspect. I work to tight deadlines. Each night I record the work done in a log for the shift supervisor. This record includes not just the work I have done but also any issues like a 'site obstruction' where I was prevented from accessing certain areas. This can happen if there is other work scheduled to take place in the same area.
These delays cause disruption and mean I will need to return to location. In my log I need to give an estimate on how long it will take to do the remaining work. The work I do helps plan the future maintenance of the tunnels.

Summary

In summary, as a technician you should know at what stage of the project you are working. You should know whether it is concept, detailed design, delivery or as-builts. Different technicians will have different experiences. Some of you will be highly specialised in the project delivery phase and may have little exposure to other stages such as project initiation. This is OK. Others may be involved in a whole range of different stages, seeing a project develop from the initial concept stage and through to design, delivery and completion. The Attribute related to 'Commercial ability' requires you to demonstrate that you have an ability to identify and organise the use of resources with consideration to cost.

Attribute 4 Health, safety and welfare

Overview

The basic purpose of health and safety in the workplace is to ensure that all workers can perform their jobs in a safe and healthy environment. Workers do not have control over their working environment, or the risks they take; employers do.

You should only carry out work if you have the relevant skills, knowledge, training and experience. You should be aware of health and safety risks involved in your work and you should understand how those risks are managed. For contractors, the use of PPE should be second nature. In every construction setting, there should be a requirement in place for some form of steel-toe cap boots, head protection and high-visibility clothing.

When working on a site, it is crucial you always follow site rules and procedures to remain safe. You have a duty to cooperate with others. In addition, if you see any hazards you have a duty to report them so they can be managed.

TPR requirements

(a)   Understand the safety implications of the role

(b)   Complete tasks with due consideration for safety

(c)   Comply with safe systems of work

This Attribute has been allocated a number and letters have been added in parentheses to assist with the explanation. These do not feature in the ICE publications.

Interpretation

Use this section of the application to demonstrate how you use your knowledge, experience and training to deliver the work you do safely. Help the reviewers appreciate the types of rules you follow and demonstrate your awareness of the types of health and safety risks that are associated with your work.

Increasingly we are asked to not just look after our physical health and safety but also the mental health and safety of ourselves and others. Many people these days will have undertaken some mental health and safety awareness training. Although there is no current legal requirement, perhaps you are a mental health first aider. In an industry with high rates of mental health issues, having an increased awareness of mental health and trained individuals can be lifesaving.

If you are a contractor, what are your duties? How do you make sure you, your colleagues and members of the public are safe during the works? For example, what procedures or guides do you use to help you set out the appropriate signage and barriers to protect a site for the duration of the works?

Do you use 'Permit to dig' procedures? What precautions do you take to avoid damaging underground services and to maintain safety on site? Do you need to undertake a CAT and Genny survey to trace underground utilities and avoid accidental strikes during digging? How do you know a scaffold is safe? Tagging a scaffold as safe to use is best practice although it is not a legal requirement. What do you or the organisation you work for do?

Working in an office as a consultant can often feel less intense when compared to working on construction sites. There is rarely a need to wear PPE in an office. But health and safety is still a huge concern. There are safety concerns relating to things like good posture and the need to understand the health risks of working with display screen equipment. It is important to look after our physical health including musculoskeletal disorders and the effects of visual fatigue, so how do you do it? It is also important to manage the risks associated with mental stress. Are risks removed or highlighted on the drawings you produce? How do you record any issues and bring them to the attention of the contractors? If you were asked to visit a site, what training would you need to do or what precautions would you need to take into consideration before visiting?

In an academic institution, what things do you need to take into consideration when managing an experiment or working with young adults? Do you work with dangerous chemicals or equipment? Are you a first aider, or are you responsible for the safe evacuation of the building in an emergency? What responsibilities do you have in laboratories, working with students and postgrads, that are different when compared to other technicians working on construction sites or in design offices?

As a client. it is necessary to have a strong attention to detail. For example, are you able to understand and comply with your responsibilities for Attribute 4(a) in relation to health and safety? You will need to know specific regulations. For example, do you know about the traffic management, working at heights, confined spaces, lifting regulations or hot work permits? Can you correctly report and maintain records in relation to your legal responsibilities? How do you deal with the general public? What processes and procedures do you comply with to be safe?

Sample responses

Contracting and construction: site supervisor nights – highway repair

At the beginning of a shift, I am responsible for conducting a briefing with the team. I set out the tasks that need to be done including any safety issues that the team need to be aware of during their work. I make sure everyone has the right PPE, is using it correctly and has the right training. I check all tools and equipment are in good working order and safe to use. I make sure temporary traffic management is set up correctly.

If working at night, it is my responsibility to make sure there is adequate lighting. At the end of the shift, I inspect the site to make sure all debris, tools and equipment have been removed and any temporary lighting has been turned off.

Consultancy and design: BIM designer – safe access for vehicles

I am responsible for checking vehicles can move safely across the site by carrying out a swept path analysis and reviewing gradient curves for various delivery vehicles. In addition I am responsible for including safety, health and environmental (SHE) boxes on my drawings. This helps us communicate health and safety challenges related to constructability issues to the contractor. I use 'hazard symbols' on my drawings to highlight specific risks the design engineer has not been able to remove. This makes it easier for the construction engineer to identify and manage issues.

Academic research: laboratory technician – general duties

I am responsible for creating a safe working environment for everyone in the lab. For example, when working in the laboratory supervising students, I am responsible for making sure they are wearing the correct PPE, such as lab coats, safety glasses, gloves and boots. I have been trained in basic first aid and I am one of the appointed first aiders.

In addition to carrying out a visual inspection of all lab equipment I am responsible for arranging PAT tests to be done on all portable tools and equipment.

Infrastructure owner/client: asset manager – wastewater drainage
I oversee daily operations of the drainage and make sure work is completed by the team safely and efficiently. I coordinate work and assign tasks. I need to make sure I keep myself, my team and any others who work alongside us safe.
I am also part of an incident response team. This means I sometimes need to respond quickly to emergencies, such as burst pipes or blocked drains, as these can pose significant health risks to the public. To be able to manage emergencies I am trained and familiar with emergency response plans that outline my role and define my duties.

Summary

You need to understand your role in relation to the work you do. You should be able to work safely and demonstrate how you can complete tasks safely and comply with the safe systems of work that have been set up. You should be able to keep yourself safe, as well as others impacted by your work.

Attribute 5 Sustainable development

Overview

Technicians are required to carry out their work in a way that contributes to sustainable development. The biggest impact we have is through our projects. Sustainability is not an abstract idea or an afterthought. It is woven into the way we approach our work.

Our industry is in a period of rapid change. Technicians, guided by standards and policies, have a contribution to make. You will need to demonstrate how you support improvements to reduce the impact your work has on the environment.

TPR requirements

(a)   Understand the principles of sustainable development and apply them in work

(b)   Complete tasks with consideration for their environmental impact

This Attribute has been allocated a number and letters have been added in parentheses to assist with the explanation. These do not feature in the ICE publications.

Interpretation

The most common definition of sustainability used is one originally defined by the UN-commissioned Brundtland Report. It defined sustainable development as that which '...meets the needs of the present without compromising the ability of future generations to meet their own needs' (Brundtland, 1987).

Members of all grades should seek training and CPD to improve their awareness and understanding of sustainability. The ICE expects every member to understand and promote ideas of sustainable development and contribute to the delivery of engineering with a balance between environmental, social and economic issues. Rule 4 of the *Code of Professional Conduct* requires us all to 'show due regard for the environment and for the sustainable management of natural resources' (ICE, 2017).

In the UK, the Engineering Council has produced the leaflet *Guidance on Sustainability for the Engineering Profession* (EC, 2021). There is benefit in downloading and reading the guide. The leaflet outlines six principles on how to achieve sustainable development in your work and meet the professional obligation to become more informed so you can play an active role as a technician.

Guidance on Attribute 5: Sustainable Development (ICE, 2021) provides advice to candidates, mentors, sponsors and reviewers on how the new Attribute 5 will be assessed at professional reviews. In particular, it looks at new requirements related to the United Nations' Sustainable Development Goals (UNSDGs) and to assessing a candidate's contribution to 'continuous improvement'.

You may or may not have heard of the United Nations Sustainable Development Goals (https://sdgs.un.org/goals). You should read up on them, but you will not be expected to quote them. Most organisations in the industry have their own systems, policies and procedures. The company procedures should be an ideal source of reference for most applicants to show how the sustainable principles apply to their work.

Another good source of information would be to consider the three pillars of sustainability: social, economic, environmental. Social sustainability focuses on the health and wellbeing of people and communities. This could also include supporting community development and education. Economic sustainability relates to more financial issues, such as creating job opportunities, ensuring work is undertaken profitably, reusing materials, or repairing and upgrading infrastructure to extend its lifespan. Environmental sustainability is about using resources wisely, minimising harm by reducing pollution and protecting natural resources for future generations. True sustainability is achieved when these three pillars are balanced and work together.

Like 'safe systems of work' (Attribute 4(c)) you will need to be able to demonstrate that you have a working knowledge of sustainability and can complete tasks with consideration for the environmental impact. In addition, does your organisation arrange corporate social responsibility days for you and your team to volunteer and help give something back? Are you encouraged to support a charity or local community group? Have you volunteered in a school to inspire the next generation?

For example, if working as a contractor, you could look to satisfy Attribute 5 by explaining how you work with others to keep noise, dust and vibrations to a minimum. What tools or equipment do you or your team use? Are there any limitations on how or when they are used or who uses them? Are you responsible for the identification and monitoring of waste materials? Do you implement your organisation's waste management plan? Are you part of an effort to use resources efficiently and minimise waste? In addition, don't be embarrassed to link the commercial savings made as an important aspect of reusing materials. What's good for the environment can often be good for the cost of the project.

If working as a consultant, you may be responsible for presenting the information in order to help the engineers and their clients make better decisions about construction or maintenance. For example, you may be repurposing buildings, reusing existing foundations or specifying low-carbon concrete on your drawing as part of a design solution. You may be supporting other more senior staff by producing carbon calculations as they link together wider sustainability requirements to benefit a project.

As a lab technician, are you responsible for monitoring, controlling and suppressing dust to prevent it escaping the laboratory? Do you need to document and manage the disposal of materials? Are you responsible for implementing energy-saving measures, such as turning off equipment? How can you evidence your responsibilities in the supporting documents?

As a client, do you consider the impact of work on the environment? Do you ensure subcontractors put measures in place to protect tree roots during excavations? Maybe you manage activities or changes to construction programmes to accommodate bats, badgers and other protected species during nesting seasons. Or perhaps you are improving bus or cycle routes to contribute to healthier, more sustainable cities.

Sample responses

Contracting and construction: concrete foreman

Where possible we aim to use proprietary formwork as this can be reused and is cheaper. During the pour I keep a close eye on deliveries to make sure concrete trucks are not standing for too long. I also keep in close contact with the supplier to check we have ordered the right amount of concrete. This helps to avoid unnecessary wastage. When striking the formwork we reuse all the materials we can or we recycle them. If this is not possible, I make sure the material is disposed of correctly in the right skips.

Sometimes we have to work within strict times or keep noise to a minimum to reduce the impact we have on local residents. I am responsible for briefing the team at the beginning of the shift and reminding them of any restrictions.

Consultancy and design: assistant CAD/BIM designer – railway arch repair

I am working with an engineer to repair an existing masonry arch bridge. The condition of the bridge has deteriorated and the line speed of the trains has been reduced. The proposed solution includes using steel arches installed below the existing bridge. My drawings help to identify the construction sequence for the work and this allows the contractor to install the steel arches at night to minimise disruption to commuter trains.

My drawings are checked and approved electronically to avoid paper wastage.

I regularly cycle, and I have taken advantage of my company's 'cycle to work' scheme and bought a new bike. This helps me reduce my daily carbon emissions.

Academic research: laboratory technician – general duties
In line with our university sustainability policy, I help ensure that e-waste is managed sustainably. I reduce energy consumption by ensuring equipment is turned off when not in use. I am responsible for making sure waste materials are properly disposed of and any material that can be recycled is processed. I use designated containers for separating and collecting concrete waste and other material like wood and metal to keep the materials clean and easier to recycle. I ensure all waste is only removed by companies with a waste carrier licence issued by the Environment Agency.

Infrastructure owner/client: asset manager – flood and coastal risk management
I work in an asset performance team. Working together we are responsible for inspecting, maintaining and repairing various assets located in river catchments and coastlines. My work includes maintaining vegetation and improving flood defences, monitoring pollution levels to check water quality and protecting and enhancing biodiversity.
I am regularly out and about in the catchment. I am responsible for reviewing various issues and pulling together projects to reduce flood risk for local residents.

Summary

Keep in mind that whatever field of engineering you work in (contracting, consulting, academia or client) you are looking to demonstrate some of the ordinary, everyday solutions you are putting in place which are an improvement on what was perhaps done historically. You are looking to demonstrate how you are involved in the sustainable development of solutions to engineering problems while giving due consideration to the impact on the environment.

Don't forget that sustainability is not just limited to the environment: it is linked to economy and social sustainability too. Often a corporate involvement inevitably will have some strategic benefit to your organisation, but it can also be personally beneficial. Volunteers often report they feel stronger and mentally and physically healthier too. If you understand that then it can help you not just personally connect with others but also do so in a sustainable way. So, when preparing your application, keep in mind all three pillars of sustainable development.

Attribute 6 Interpersonal skills and communication

Overview

Good communication skills are important. This Attribute should not be underestimated. It's not just about how you tell people what to do. It's more than that. It's how we let other people know what we are thinking and feeling and how we get to know what others are thinking and feeling. It involves verbal and electronic communication, both how we say it and how we listen to others.

How do you communicate in your team? Do you attend regular team meetings? Are these in person or online? How do you record what work you are doing or what work needs to be done? Do you have a programme of work? Are you responsible for recording daily tasks or progress made on any issues? Do you need to create or monitor project logs so you (and your team) are aware of how a project is progressing, including what you delivered and what resources were used?

How do you work with others? What do you mean when you say you 'respect others' and how do you encourage people to work together? On the day you will have the opportunity to communicate by doing a five-minute presentation and interview. Alongside your application, these will also help you demonstrate how you communicate.

TPR requirements

(a)   Communicate effectively with others, at all levels, in English*

(b)   Work effectively with colleagues, clients, suppliers or the public

(c)   Demonstrate personal and social skills

(d)   Demonstrate awareness of diversity and inclusion

Please note: All assessments and reviews for Engineering Council registration will be conducted in English, subject to the provisions of the Welsh Language Act 1993

This Attribute has been allocated a number and letters have been added in parentheses to assist with the explanation. These do not feature in the ICE publications.

Interpretation

Depending on your role, you will find there is a lot of overlap in the roles undertaken by infrastructure owners and clients compared with consultants and contractors. Think about how you communicate.

For example, if working as a foreman or supervisor, you are the link between the construction workers and the site managers. You will have a key role in taking responsibility for overseeing work gangs and ensuring tasks are completed efficiently, safely and to the required standards. If you ordinarily work on the highway, for example, then your works are likely to be on or near a footway where there is a risk that pedestrians might enter the working space. Think about some of the less obvious forms of communication, such as site signage and notice boards. How do you communicate with the public?

At the beginning of your shift, what do you do? How do you know what needs doing and how do you share this information? How do you make others aware of the associated risks and safety measures that have been put in place? How do you ensure everyone knows their roles? What records do you keep to confirm that everyone has been briefed and is ready to work? When you leave a site, do you just simply go home, or do you have to provide a report to a project manager to summarise the work done and any issues that need attention?

Consultants often communicate through their drawings. This is how complex information is presented to contractors and clients. If you are a CAD/BIM technician, what is your role? How do you communicate with your engineer? How do you translate their information into a model or drawing? How do you make sure the information is easily understood? How do you know what you have to do each week? How is that communicated to you? What records do you need to keep of the work you do? Do you change the way you communicate depending on who you are talking to? Do you visit schools or work with students on work experience?

If you work in a university's laboratories, how does your management communicate with you? How do you know what groups will be using the labs? How do you know what they will be doing and what they will need? When explaining how equipment works, how do you communicate with the students? Is this different from how you communicate with a postdoctoral student? Are there any guidelines for communication with vulnerable adults which impact on your role?

If you are working as a streetworks inspector, you may need to communicate with members of the public, utility companies and their contractors. Working alongside an engineer, you may be responsible for producing plans and proposals for local councillors or members of the public. How do you do it? What evidence could you include in your application? What different software packages do you use to communicate with other people? How do you use your ability to communicate to deal with complaints and compliments from different stakeholders?

> "Professional technicians are people who take the time to stop and pause, check their work and use their knowledge and experience to eliminate errors from their models and drawings.
>
> They are people who fundamentally understand how drawings go together. They get into the feasibility of the design details by applying their knowledge and experience. They do not get lost in the assurance process. They stop and pause and, to put it simply, take pride in the quality of their work."
>
> **Tom Price EngTech FICE ACIArb**

Think about how you ensure everyone feels respected. What is your responsibility? How do you help someone settle in when they are new to the team? We all have the right to not be discriminated against simply because of who we are. Some countries have laws to protect people from discrimination.

In the UK, for example, it is against the law to discriminate against age, disability, gender reassignment, marriage and civil partnership, pregnancy and maternity, race, religion or belief, sex and sexual orientation. Together these are known as 'protected characteristics'. Diversity and inclusion are core values at the ICE. When we work with many different people we connect better. More diversity means more breadth and depth of ideas.

However, you should be able to demonstrate your personal skills and awareness of diversity and inclusion without needing to refer to the local law. Think about the team you work with. You will quickly find you work with people who are different from you. They may be older or younger. They may be single, married or divorced. There will be people of different races, faiths, religions or ethnic backgrounds.

Sample responses

Contracting and construction: concrete foreman – 'site person in charge'

Good communication is essential to ensure a smooth, efficient and safe site. I am responsible for carrying out a site briefing at the beginning of the shift before any work is started. Once complete I ask questions to make sure people have taken on board what they have been told. At the end of the briefing, I make a register of who was present.

At the end of my shift, I make sure I complete a detailed daily diary sheet of the work done. I need to record the key items, such as date, weather conditions, the number of workers and different trades involved. These records are for the quantity surveyor.

Consultancy and design: assistant CAD/BIM technician

I work with CAD software MicroStation to develop 2D and 3D models for various tasks, including typical details and track drainage. I then create detailed drawings from these models to enable the client to see our design intent. I decide on what is an appropriate scale for details and the number of drawings to use.

I make sure all drawings are named using the correct convention. If there are changes to a drawing I revise and amend drawings and models according to mark-ups produced by engineers. I then update the revision number to provide a clear record of changes made and to help identify who made specific changes and when they were made.

Academic research: laboratory technician – general duties

The clear communication of information between staff and senior management is vital as this ensures everyone has the necessary knowledge to do their jobs. This is especially important at the beginning of the academic year when there are a large number of new students coming to the university for the first time.

When working in the labs we take great care to communicate risks to students. We also need to take into account safeguarding issues for those students who are under 18 years old.

Infrastructure owner/client: asset manager – highways

I am responsible for the day-to-day management of work on the highway. My team carries out routine maintenance, such as repairing potholes, clearing drains or maintaining road surfaces. At the beginning of each shift, I need to check the task list for the day and check we have the right materials, tools and equipment. I then give a quick briefing to the gang to make sure everyone is clear on their roles and we get started with the work.

Once the work is done, I take photographs to visually document the work done. I work with a variety of people from all sorts of different backgrounds. I work hard to make sure they respect each other and work as a team.

Summary

Tackling this Attribute can be far simpler than it first appears. As you prepare your application, think about all the different forms of communication you use and how can you provide some evidence of them. Consider who you communicate with and how you communicate with them. Effective communication can be about building good professional relationships between different people and respecting differences.

Attribute 7 Professional commitment

Overview

Professional commitment is crucial. If you take pride in your work and commit to producing high-quality results, you will create a positive professional working environment. But how do you demonstrate your commitment? Is it just a case of doing some training and following the rules? Or is it much deeper and more meaningful that that?

This is not simply an add-on Attribute; it is at the very heart of what it means to be a professional member of the ICE. It could be argued that the process of registration requires a professional commitment and therefore automatically offers a means to demonstrate this Attribute. But a professional commitment is more than simply preparing for a review and passing it.

By caring about the quality of your work, by making a commitment to stay up to date with industry trends and best practice and by being a reliable and trustworthy member of the team, you will be doing the right thing. Integrity and honesty are the very foundation of what makes someone a professional. To be successful in your TPR you need to be able to demonstrate how you behave like a professional.

TPR requirements

(a)   Understand and comply with the ICE Code of Conduct

(b)   Understand the ethical issues that may arise in their role and carry out their responsibilities in an ethical manner

(c)   Carry out and record the continuing professional development (CPD) necessary to maintain and enhance competence in their own area of practice

This Attribute has been allocated a number and letters have been added in parentheses to assist with the explanation. These do not feature in the ICE publications.

Interpretation

All the requirements of this Attribute could be viewed as being interconnected with the *ICE Royal Charter* (ICE, 2023a) which is accessible online (https://www.ice.org.uk/download-centre/royal-charter-and-by-laws). For example, your ability to work ethically is linked to your knowledge and understanding of the *ICE Code of Professional Conduct* (part 7(a), ICE, 2017). The code itself has six rules and is easily found online (https://www.ice.org.uk/download-centre/code-of-conduct). Questions on the code are commonly raised by reviewers and you should arrive with a good understanding of the rules and how they relate to you and your experiences.

Summary of the six rules of professional competence

1. All members shall discharge their professional duties with integrity and shall behave with integrity in relation to all conduct bearing upon the standing, reputation and dignity of the Institution and of the profession of civil engineering.
2. All members shall only undertake work that they are competent to do.
3. All members shall have full regard for the public interest, particularly in relation to matters of health and safety, and in relation to the well-being of future generations.
4. All members shall show due regard for the environment and for the sustainable management of natural resources.
5. All members shall develop their professional knowledge, skills and competence on a continuing basis and shall give all reasonable assistance to further the education, training and continuing professional development of others.
6. Members must promptly inform the Institution of serious criminal convictions, disciplinary terminations, bankruptcy or significant breaches of conduct by others. They should also report any risks or wrongdoing to employers if protected by law and support colleagues who raise such concerns.

"When I found out I had passed the EngTech MICE there was a lot of relief and a bit of pride. I understand the value I bring to projects, but that's not always noticed by others. It was a really nice feeling to have my skills recognised by the ICE."

George Woods EngTech MICE

Ethical issues in Attribute 7(b) are less easy to define. The Engineering Council and the Royal Academy of Engineering (RAEng) have put together an excellent guide to professional practice and behaviour. It is worth downloading a copy (https://www.engc.org.uk/media/2334/ethical-statement-2017.pdf) (RAE, 2017).

What are your responsibilities for safety, reporting 'near misses' or control of pollution? How is waste handled and removed? What should you do, or not do, in respect of fraud, gifts, hospitality and so on? Ultimately, ethics are about honesty and integrity: knowing what to do, being honest, treating others fairly, taking responsibility for your choices and doing the right thing at the right time.

Increasingly, technicians are getting involved in the promotion of the industry; for example, you may be supporting STEM activities by going along to a local school. So, if you are involved in supporting apprentices, school leavers or members of the public visiting your sites or offices, this all adds to building evidence of your commitment to yourself and your profession. It also demonstrates how you are assisting others in developing their career.

Being able to identify the limits of your own knowledge and skills is important. It is so important that it is defined in Rule 2 of the ICE *Code* (ICE, 2017) which says, 'All members shall only undertake work that they are competent to do.' This is why knowing your limits is not just common sense but an extremely important aspect of your role in the industry. In the TPR you should explain where you have done this. How do you assess your skills? Who helps you do this? How do you seek improvements or develop your knowledge?

Continuing professional development (CPD) is identified in Attribute 7(c) and is another opportunity for you to demonstrate your professional commitment. This can simply be done by submitting a development action plan (DAP) and a professional development record (PDR). If you do not submit a valid, current DAP or your PDR does not cover at least one year and is not up to date then your application is likely to be rejected. The approach to recording CPD is one that is essential for the preparation for the TPR. This indicates that you are committed to maintaining and developing your skills and you have undertaken CPD to continuously strengthen your experience and capability. Developing your skills and knowledge is a continuous process. In the TPR the reviewers will also want to understand your commitment to lifelong learning.

Sample responses

Contracting and construction: storeman
I maintain and manage the inventory of materials, tools and equipment. I make sure everything is accounted for by keeping detailed records of all materials and equipment, including the condition and usage. I monitor stock levels for different materials and make sure they are stored safely and securely to prevent theft, damage or accidents.
When we take a delivery, I take responsibility for checking it against the orders. If there are any inconsistences or errors, I notify the supplier. By doing this I can make sure the problem is quickly sorted out. Audits are carried out by the site manager to check I am carrying out my role correctly and efficiently.

Consultancy and design: CAD technician
I must accurately record the time I spend working on projects. I usually fill my timesheets in at the end of the day or immediately after completing an element of work. I need to keep myself up to date with the latest CAD software and be aware of best practice to help maintain my abilities. There is always something new to learn.
I believe it's really important to pay attention to the details. I check my own work and follow company guidelines too. This helps prevent errors.

Academic research: laboratory technician – general duties

Before I use a new piece of equipment, I reach out to someone more experienced than me and ask them for guidance and support. Sometimes this is not enough, and more formal training is required to make sure I can use the equipment safely.

In my role I often work with groups of students. This can include people under the age of 18. I have undergone safeguarding training. This helps me create a safe and supportive environment and makes sure I am meeting legal obligations and best practice.

Infrastructure owner/client: highway network – inspector

I carry out routine hazard and safety inspections to ensure the safety of the public is not affected by the condition of the highway. If I identify an emergency, I take responsibility to rectify it within 2 hours. For less urgent defects, I notify the maintenance team who will seek to repair the defect within 4 weeks.

I frequently interact with the general public. I have been given training to improve my ability to communicate. I always seek to treat people with courtesy and respect. Sometimes this can be very difficult, but I believe in being proactive, open and honest. It helps me to carry out my work and it helps build trust.

Summary

As a professional, you have a duty to always act with diligence and care and keep your knowledge and skills up to date. You could also consider how you assist others in developing the skills they need. Ethics are about how you treat people. You should (as discussed in Attribute 6) promote equality, diversity and inclusion. Being reliable and trustworthy involves a respect for life, law, the environment and the public good.

By providing a copy of your CPD records you will go a long way to demonstrating your commitment, but they are not the complete picture. To be able to work as a professional technician takes time and commitment. It requires energy and enthusiasm to get to know how to do a job well. It is something that unfolds over time, requiring dedication and a commitment to continuous learning. This is why it is necessary to demonstrate your professional commitment at the TPR and show where you have applied your knowledge.

Case studies

The following pages provide a couple of case studies which summarise the types of evidence that could be used to cover the full range of the Attributes. As a candidate, you should reflect on them and discuss them with your mentor(s). With their help you will develop your understanding of how to use your experiences to demonstrate the Attributes yourself.

Case study 1 – Foreman's role in a concrete pour

Attribute 1 Understanding and practical application of engineering

If responsible for supervising the construction of concrete pour, initially you will make sure the setting out for the formwork is correct. You may have to check the formwork and that it is supported as designed. On a large project someone else may fix the reinforcement but you may be responsible for checking it's correct. It may be your role to coordinate the delivery of concrete and check it arrives on time and is the correct mix and quality.

During the pour you will make sure the concrete is vibrated and compacted, monitor formwork and any designated pour rates. Once the pour is complete it will be your role to supervise the levelling and finishing. Through your experience you will know how hard it is to achieve a 'special' finish compared to a 'basic' finish. You will then ensure the right tools, equipment or fine abrasives are used.

Attribute 2 Management and leadership

You will be responsible for making sure everyone in the gang knows what to do and is competent. You will check through what needs to be done and make sure you have the time to do it and the right team in place. You will check you have the right tools and equipment and that they are in good working order. You will make sure the concrete is placed and spread correctly.

At the end you will conduct a final inspection to check the finish and compile records. You will log all the relevant information as part of a quality control procedure.

Attribute 3 Commercial ability

During the pour, you will make sure everything goes smoothly. You will keep an eye on the weather and record progress. You will coordinate the concrete deliveries, manage a gang or subcontractor with the aim of completing the pour on time with minimal waste.

You will be responsible for maintaining accurate records of delivery tickets and test results. You will know what to do if the concrete fails a slump test or what happens if anything delays work, such as bad weather or late deliveries. You will record all this in your daily log (or site diary).

Attribute 4 Health, safety and welfare

You will check that the gang are familiar with the work they need to do and how to do it. You will ensure the team have safe access to and egress from all stages of the pour. You will be responsible for checking whether they have the right training, plant and tools and that appropriate PPE is worn. As part of your responsibilities, you may also check that the formwork is stable and a 'permit to load' is in place.

Prior to the pour you may gather the gang together to give a safety briefing or 'toolbox talk', after which you will check everyone has understood and confirms their attendance. During the pour you will make sure everyone works safely. You will know what to do if there is a reportable incident.

Attribute 5 Sustainable development

You may be working within strict noise, dust or vibration limits. You will know what to do with surplus concrete. You may be responsible for making sure wash-out and wheel wash facilities are used. You will know how to reuse materials or how to dispose of them.

Attribute 6 Interpersonal skills and communication

You will use a variety of ways to communicate. This could include emails, phone calls and site sketches. You may be responsible for toolbox talks, making special allowance for those for whom English is not their first language. You will keep formal records of some activities, such as diary sheets, permits to pour and permits to load, as a permanent record. You will treat all people with respect. You will be able to take responsibility for a gang's diverse characteristics, such as gender, age or nationality.

Attribute 7 Professional commitment

You will make sure you and your gang have the right qualifications, training and experience to do the work you do. You will set a good example and take responsibility for your work. You will treat people with respect and fairness. You will be trustworthy.

You will keep a record of your training and know the limits of your responsibility. You will appreciate how the ICE Code of Conduct applies to you and your work.

Case study 2 – CAD technician producing drawings for a new bridge

Attribute 1 Understanding and practical application of engineering

You will be familiar with the software you use to draw or model the engineer's sketches for a bridge design. You will plot the highway alignment and may use this to set out the geometry from the bridge. You will use appropriate scales for your drawings to illustrate general arrangements, cross-sections and connection details. You will be familiar with engineering terminology, such as parapets, bearings and abutments.

You will provide suitable information for the construction of the bridge, including dimensions, setting out points and annotation covering construction elements (i.e. construction joints), materials and tolerances. You will create the drawing in the right coordinates, referencing survey records and following any client or company requirements, such as drawing/modelling standards or use of a common data environment.

Attribute 2 Management and leadership

You will plan your workload to ensure you can also deliver drawing packages to project deadlines. You will work as part of a team. You may coordinate resources or provide mentoring to other technicians. You will work alongside engineers to deliver technical drawings that are clear, well-presented and are independently checked. Perhaps you quality check drawings produced by others.

You will have a keen eye for detail. You will make sure your drawings are well laid out and clearly labelled. You will follow robust procedures for managing, reviewing and exchanging drawings and models between your company and the client.

Attribute 3 Commercial ability

You will understand the purpose of the drawings you are producing – for example, for a 'tender' bridge design or 'construction issue'. You will ensure your drawings are of an appropriate detail, quality and suitably labelled for the purpose they are issued. You will complete your timesheets and take care to record the hours you have worked on individual projects. You will know whether additional work is being done outside of the original scope and you will record your time separately. You will understand the cost of your time and how this relates to project or business costs.

Attribute 4 Health, safety and welfare

Where possible you will work with the engineer to eliminate any hazards in the bridge design using CAD software. Where this is not possible you will seek to reduce them by highlighting hazards on the drawings so the contractor can isolate them or control the risk.

You will ensure you work safely by checking your display screen equipment. You will take regular breaks. In addition to your work, you may be a fire warden, first aider or mental health first aider.

Attribute 5 Sustainable development

You will be aware of the impact your work can have to improve the lives of people, such as those who need to cross the railway safely. By eliminating stairs or providing alternative means of access you will ensure the design you draw is accessible to a broad range of users. You will minimise your carbon footprint and, where possible, seek out efficient ways to travel to work, such as taking advantage of a cycle to work scheme. You may use electronic printing and review processes to avoid using paper.

Attribute 6 Interpersonal skills and communication

You will lay out your drawings so the contractor can understand the construction sequence and build the proposed design. You will communicate clearly at the appropriate detail and scale the information required.

In addition to your normal work, you may be asked to support a schoolchild who is on a work experience week. You may work with people who have a different belief, race or religion from you.

Attribute 7 Professional commitment

You will care about the work you do and take pride in doing it well. You will act responsibly and set an example for others. You will seek to maintain your skills and, where appropriate, develop them in new areas. You will keep suitable CPD records. You will be self-aware of your level of experience and what work you are competent to carry out.

REFERENCES

Brundtland G (1987) *Report of the World Commission on Environment and Development: Our Common Future.* United Nations General Assembly document A/42/427.

EC (Engineering Council) (2020) *The UK Standard for the Professional Engineering Competence and Commitment (UK-SPEC)*, 4th edn. EC, London, UK.

EC (2021) *Guidance on Sustainability for the Engineering Profession.* EC, London UK. www.engc.org.uk/sustainability-leaflet (accessed 30/03/2024).

ICE (2017) *ICE Code of Professional Conduct.* ICE, London, UK.

ICE (2021) *Guidance on Attribute 5: Sustainable Development, Version 1, Revision 0.* ICE, London, UK.

ICE (2022) *Construction Site Engineering Technician Level 4 Mapping of Knowledge, Skills and Behaviours Against EngTech MICE Attributes, Version 2, Revision 1.* ICE, London, UK.

ICE (2023a) *Royal Charter and By-Laws.* ICE, London, UK.

ICE (2023b) *Technician Professional Review Guidance, Version 3, Revision 7.* ICE, London, UK.

RAE (Royal Academy of Engineering) (2017) *Statement of Ethical Principles for the Engineering Profession.* RAE & EC, London, UK.

Successful Professional Reviews for Civil Engineering Technicians

Malcolm Peake
ISBN 978-1-83549-943-6
https://doi.org/10.1108/978-1-83549-940-520251005
Copyright © 2025 by Malcolm Peake. Published under exclusive licence by Emerald Publishing Limited

Chapter 5
Putting the submission together

Overview

Once you have decided you want to be recognised as a professional technician, the next obvious question is, 'How do I apply?' The ICE offers a lot of guidance online and there are some excellent workshops, which are available both online and in person. Initially things will look relatively simple, but the more you investigate the process, the more daunting it might appear.

This chapter will give you a step-by-step guide on how to put your submission together for the direct EngTech MICE application for the technician professional review (TPR). If you are applying for EngTech MICE as part of your apprenticeship, then this chapter will be helpful. Further details are provided in chapter 7 which are more specific to the end point assessment (EPA).

> "EngTech MICE proves that you are a valuable asset to any team. It will give you opportunities outside of those you may expect as your career progresses due to the recognition from your superiors that you are willing to go that extra step!"
>
> **Tom Price EngTech FICE ACIArb**

Before preparing your TPR application, it is crucial to have a clear understanding of what the technician member grade entails. The ICE requires technicians to be able to 'apply proven techniques and procedures to the solution of practical engineering challenges'. Therefore, familiarise yourself with the Attributes, competencies and roles expected at this level to ensure that your application is in line with the expectations set by the ICE.

Did you know?

The Jean Venables Medal is presented every year to recognise and celebrate the best performance by a newly qualified Technician Member at a TPR in that year. Reviewers nominate candidates who have shown a passion and enthusiasm for civil engineering, demonstrated active involvement in ICE activities and displayed excellence across all Attributes.

This is a unique, prestigious award recognising enthusiastic and outstanding technicians.

By following the steps in this chapter and working with your sponsor(s) you will be giving yourself the best opportunity to demonstrate your knowledge, experience and abilities and successfully be awarded EngTech MICE.

"I've been working as a civil engineering contractor for the last five years. Since I started, it's been a dream of mine to become chartered. The EngTech application and accreditation was the first step. The entire process gave me a glimpse of being a chartered engineer as it resulted in recognition and affirmation of my ability as an engineer."

Ibrahim Kapasi EngTech MICE

The application form should not be rushed. There is no excuse for last-minute submissions given that the ICE publishes dates of reviews a long way in advance; typically, over six months ahead.

As you become more familiar with the process, you will realise you need to gather various pieces of documentation and evidence to support your application. You will need to ensure you have relevant records of your qualifications, your work experience and any training and courses you have completed. You should develop a portfolio that includes material that demonstrates what you contribute to different civil engineering projects. This chapter will go into more detail to help you appreciate how you do that.

The aim of the application is not to tell the ICE about everything you ever did. It is more along the lines of whether you are a professional and able to do a good quality job and whether you can use proven techniques and procedures to solve practical engineering problems. Therefore, your application needs to demonstrate this through the projects and examples you choose to write about and the evidence you choose to provide. This is why it is no good copying someone else's submission; it simply will not work. You are an individual; what you need to do is ensure that *your* picture is complete.

When putting together your application it is essential to seek guidance and mentoring from experienced professionals who understand the application process. This can greatly enhance your preparation for the TPR. As mentioned in chapter 3, you will need two people to act as your sponsors and they will need to be familiar with your work. You are also encouraged to connect with other civil engineering technicians in your company who have already undergone the process. They will inevitably share their own experiences and provide some helpful tips to boost your chances of success.

If you work in a small company, then reach out to the ICE and take the opportunity to network with people in your local area, either online or in person. If you find it easier to discuss things and talk them through rather than just reading guidance, then you will find there is an abundance of opportunities to get guidance on the review process.

Did you know?

There are many existing members of the ICE who will happily provide valuable insights and advice. If you are not sure where to start, try searching up 'ICE Technician Tea Break' or 'ICE Technician Timeout' and book yourself into one of the two virtual meetings held almost every month.

All the elements of the application fit together like the pieces of a jigsaw. Once complete, this will be your picture of you. The way the pieces fit together is similar for everybody at the TPR. However, the individual pieces are different shapes and sizes, reflecting each individual applicant. No two jigsaws will ever have the same shape or size pieces. In this chapter we will look at how you can set about building your jigsaw and what it could look like.

Image courtesy of Harry MacDonald Steels taken from Waterhouse, P (2022)

Apprentices

The overall purpose of application for the EngTech MICE has many similarities with the level 3 and level 4 end point assessments, but the details associated with the application process and interview day are very different. For example, the EPA has two parts to the assessment day. They bring together a combination of the academic knowledge gained in the classroom and the industrial knowledge gained through work. The assessment can include an EngTech MICE award.

The character and capabilities sought by the assessors in either a level 3 or level 4 EPA assessment have much in common with those sought by the reviewers in the TPR. There is a common theme and a natural connection between the EPA assessment and the TPR. However, it is important to understand the subtle differences. This chapter will focus on the Attributes and the TPR. You can find out more about how they link to the apprenticeship standards in chapter 7.

Sponsor's statement of support

As a brief aside before we look at the main details of the application, it is worth taking time to consider the sponsor's statement of support (ICE, 2023a). All applications for membership must be supported by *two* sponsors. You are not responsible for preparing the statements of support. These are written independently by the sponsors.

Did you know?

Sponsors must meet certain requirements, so it's important that you read up on who is eligible to be a sponsor and what they are required to do. If you want to know more then read chapter 3.

Ideally, you need to locate a good pair of sponsors. They will be invaluable. The ICE has a dim view when they receive poor submissions which are incomplete, badly laid out or have missing information. With the right support and attention, you will feel more prepared, less nervous and ready to do your best on the day. The reviewers expect to meet a professional, perhaps a slightly nervous one, but a professional all the same.

Your sponsors should be familiar with the requirements of the route you are following (e.g. TPR or EPA) and not fall back on their own recollections of when they passed their review. It will be beneficial if both *you* and *your sponsors* are familiar with the whole process. You will be in a better position to demonstrate your abilities if you fully understand the process. Therefore, allow adequate time to prepare your application.

Did you know?

Both sponsors must upload their statements *one week* before you apply for your TPR; check online for more details (https://www.ice.org.uk/join-ice/key-membership-dates).

The application form

The first thing to do is make sure you are looking at the latest *Technician Professional Review Application Form* (ICE, 2024b). The Institution has set out its requirements for the application in its publication *Technician Professional Review Guidance* (ICE, 2023c). As you look through the form, in addition to simple questions about your name, address and other personal details, you will find other questions related to presentations, CPD records, Attributes and appendices.

"Initially I felt that it looked like a lot of work and resembled a job application on steroids. After reading through it a few times and digesting it, I got a general feeling of how I'd progress with it. I chipped away at it over a few days, so it was easy enough in the end."

Rob Ehren EngTech MICE

It is an unusual thing to write about yourself and is quite likely to be an unfamiliar thing for most applicants. It may be easier to think of your submission more like a portfolio of evidence which you can use to demonstrate the type of work you can do and how you do it.

The *Technician Professional Review Application Form* (ICE, 2024b) is a standard-looking form which needs to be completed and submitted to the ICE. At first glance it all looks very sensible and there does not appear to be anything too complex. Don't be tempted to leave this to the last minute.

The form needs care and attention, so take your time to complete it to avoid any unnecessary errors. Carefully review the application process outlined by the ICE for the TPR. Take time to understand the specific requirements, deadlines and supporting documents needed for a successful application. Be sure to gather any evidence required to demonstrate your competence for the appendices. The form is split into six sections; we will look a little closer at what each section needs.

Section 1 – about you

The first piece of information you are asked to provide in section 1 of the application is your ICE membership number. If you have already registered with the ICE as a student member then this is a relatively simple question. However, this question often stops many people in their tracks. How do you provide a membership number for an organisation of which you are not yet a member? Although at first glance this question appears to be unanswerable, the answer is far simpler than it first appears and there is even a useful link in the ICE guidance which will send you to https://myice.ice.org.uk/register.

By following the link, you will be taken to a web page where in a matter of minutes you can register for an account with the ICE and it will generate a unique eight digit personal identification number. This will be the number you will use in your application. After this there are the more obvious questions to answer and information to include, such as name, date of birth, home address, email address, nationality and gender.

You are required to include a recent photograph on the front cover. There is a space for you to attach a photo. A picture of you in hard hat and fluorescent jacket may show you at work, but it is not always useful in identifying you as the person at the interview. The photo is used for identification purposes, so it is recommended to pick a good quality image. The photo must be a true and current likeness of you. Choose one that has a clear image of you and consider one that has a solid, plain background. If you follow the guidance for a standard passport-style photograph, you will not go far wrong.

Qualifications

Later in this section, you will be asked to provide other information relating to your employer and your qualifications. These are all standard questions and should not trouble most people. Most applicants will have a relevant qualification, such as a level 3 NVQ in Highways Maintenance or an HNC in Civil Engineering.

For those of you who have already got the appropriate qualifications which mean you are eligible for EngTech MICE, you should list them in the table provided and include a copy of them as part of your application (see section 4). If you are unsure whether your qualifications are eligible, you can check the list of approved technician courses at https://www.jbm.org.uk/ or contact the ICE directly.

Data on all accreditation activities is held at the ICE so if in doubt, check out which grade of membership you are eligible for at https://www.ice.org.uk/join-ice/accredited-course-search. The ICE will not process your application until they have received all the relevant information. Therefore, it is best to start this process early to avoid disappointment and delay.

However, the sharp eyed among you will have noticed you can apply with or without approved qualifications. If you are not able to provide evidence of formal qualifications, all is not lost. You will be given the opportunity in the TPR to demonstrate that the experience and knowledge you do have are equivalent to anything you could have learnt at college. What you need to do is complete the form to say you will be applying for a TPR 'without approved academic qualifications'.

Did you know?

If you are applying without approved qualifications, you must be able to demonstrate both your understanding of the engineering principles and knowledge of how to apply them. To enable you to do this, your interview is extended from 45 to 60 minutes. The additional time is to allow for a primary focus on your technical knowledge. There is more information on this in chapter 4 and chapter 6.

Language

Most professional reviews at the ICE are conducted in English (or Welsh if applicable). To be eligible for registration with the Engineering Council you must be capable of conducting your TPR in English, Welsh or British Sign Language (BSL).

Unspent convictions

It is quite clear the ICE will not admit any person with an unspent conviction relating to a serious criminal offence to any grade of membership, unless there are special circumstances that show beyond reasonable doubt that the person is a fit and proper person to be admitted to membership of the Institution. And make no mistake, this is for 'serious criminal offences'. This relates to any offence punishable by a court of competent jurisdiction by a term of imprisonment of 12 months or more.

If you do have a criminal conviction and you are considering applying for membership of the ICE, then to progress your application you must complete the *Unspent Convictions Form* https://www.ice.org.uk/download-centre/unspent-convictions-form (ICE, 2023b). It must be signed by both sponsors and submitted with your application. Your application will be held until permission is given to proceed. Your application will be dealt with sensitively and confidentially by the ICE. The information will be held securely by the Professional Conduct Team.

Career history

The final item of section 1 asks for your career history. This is essentially a summary of your CV, starting with details of your current employer. You should list your employment history, including the job titles and start and end dates of each employment, and include a brief summary of your roles and responsibilities. All this needs to be completed within *one page*.

Section 2 – professional review requirements

It is in section 2 that things start to get exciting. It is here you need to decide if you want to take the opportunity to give a five-minute presentation. It is where you decide on when and how you would prefer to sit your TPR. It is where you will specify your area of technical expertise and employment type and put forward the names of the people who have agreed to sponsor your application.

The five-minute presentation is optional. It is important to remember this: for some people the desire to just get on with the interview will be far more appealing than the thought of doing a presentation. This is fully understandable. However, it is fully recommended you take the opportunity to do a presentation. You can find out more about preparing the presentation later in this chapter and delivering it in chapter 6.

You need to decide when and where you would prefer to be reviewed. You should check out the ICE 'key dates' on the website and choose a suitable review session. These usually take place three times a year (February, June and October). Please note that reviews can be held in person or online. So, you need to identify not only your preferred session but also your preference for review, whether online or in person. You will find more information to help you decide in chapter 6.

You also identify your area of technical expertise and employment type (see Table 5.1 below). Keep in mind that the ICE will use this information to automatically match you to one of your reviewers. Therefore, it's better to select the area that best matches the work you are submitting.

Finally, for this section, there is the option to choose whether you would like to have your name published on a list of successful candidates. Many of you will be proud to be recognised. But some of you will wish to keep these things private. If you wish to remain anonymous then do not tick the box at the bottom of section 2 and the ICE will respect your wishes.

Table 5.1 Engineering expertise and employment types

A. Technical expertise (select only **one**)

Bridges	☐	Dams / Reservoirs	☐
Environmental planning / engineering	☐	Building, Structures	☐
Airports	☐	Water Supply / Sewerage Treatment / Drainage and networks	☐
Geology, Geotechnical and ground engineering, Tunnelling	☐	Railway Systems and infrastructure	☐
Offshore Engineering	☐	River, Coast, Marine, Docks and Harbours	☐
Transportation Planning	☐	Highways & Traffic Engineering	☐
Regeneration and Development	☐	Energy services	☐

B. Employment type (select only **one**)

Contracting and Construction	☐	Consultancy and Design	☐
Academic Research	☐	Infrastructure Owner / Client	☐

C. Infrastructure engineering description (infrastructure candidates only)

Building services	☐	Production and manufacturing	☐
Chemical, process (and energy)	☐	Materials	☐
Digital / telecommunications	☐	Mechanical	☐
Electronic and electrical	☐		

All infrastructure engineer applicants note and summarise engineering specialism (in up to 200 words)

(ICE, 2024)

Section 3 – demonstration of the Attributes

This is where you write a report about yourself. Rather than ask you to start with a blank piece of paper, the ICE has identified the seven Attributes of a professional. These are covered in detail in chapter 4 of this book. The key thing to keep in mind is that you do not necessarily have to spread the 3000 words equally across the Attributes.

A key aspect of the TPR is demonstrating your technical competence. Therefore, it is important that you identify the Attributes outlined by the ICE and show how they align with your area of expertise. For example, the technical skills of someone working in an engineering consultancy are very different from those of a supervisor on a construction site (ICE (KES), 2023).

> "With regards to the Attributes, my advice is always the same: only write in your application subject matters you are confident talking about."
>
> **Aaron Passfield EngTech MICE (TPR Reviewer)**

You should focus your application on one or two engineering challenges you have been involved in resolving. If you are applying without approved qualifications, you should write more about the technical side of your role and take the opportunity to develop this deeper before you highlight some of the quality, safety, management, commercial and environmental issues you need to address as part of your work.

The project(s) or piece of infrastructure chosen as the background for the Attributes will vary widely between one application and the next and it is for this reason chapter 4 of this book has been entirely devoted to the Attributes.

Section 4 – supporting documents

Together with your completed application you must include the following supporting documents. These are often referred to collectively as the appendices or supporting evidence. The terms are all interchangeable. You must include the following as part of your pdf submission:

- a recent photograph
- a certified copy of your qualifications
- appendices illustrating your work
- continuing professional development records.

Photograph

The ICE reviewers need a recent photograph of you to help them recognise you on the day of your interview. There is space on page one of the form to attach a photo.

Certified qualifications

Once you have established that you are eligible for EngTech MICE you need to arrange for all your qualifications to be certified and saved as pdf files ready to be attached to your application. Each certificate should be signed by a professional person. Ordinarily people ask one of their sponsors to do this, but you can also ask a tutor or any senior member of your employing company.

Did you know?

The person signing the copy of the certificate should write the following:

"I confirm this to be a true copy of this applicant's qualification".

Signature:

Print name:

Employing organisation/college:

Position:

Contact telephone number or email:

ICE membership number (if applicable):

The guidance states that 'it is important to include examples of your work, as they help the reviewers assess your competence' (ICE, 2023c). The aim is for you to provide supporting evidence that assists in demonstrating the abilities you have highlighted in the Attributes.

Did you know?

In the appendices you can submit:

- three A3 documents or drawings and
- twelve discrete A4 pages of additional information.

The appendices are not included in the word count.

The appendices are nothing more than supporting evidence that illustrates your work and proves that you have done what you claim you have done (see Figure 5.1). The 'perfect' evidence will have a project name, your name and the date you were involved written on it. This is ideal for some evidence, such as drawings, emails, diary sheets or any document that forms part of a quality process.

Another way of thinking of the supporting documents is to look at building a portfolio of evidence of what you have recently been working on. Think about what you need to do in your ordinary job. How do you prove you did a test on site? How do you demonstrate you have carried out a check or you have followed a process? What do you sign, what do you put your name on? What do other people put your name on? All these things will be evidence that you are doing work and taking responsibility for the work you (and sometime others) do.

Don't just think about the product of your work. There will be evidence of others putting your name on things, such as resource charts. If it is not clear how the evidence relates to you and your role, consider including a short sentence to clearly state how it is relevant to you.

Carefully consider each piece of evidence you choose: does it add value? Does it help prove to the reviewers you are a professional technician? Does it show them what you can do or need to do to make sure your work is done to a high standard? Any evidence you submit is open to scrutiny so be prepared to discuss it and answer questions about it in more detail.

If you want to show someone you can draw or model and you need to do check prints to make sure you communicate with others in your office, then include a copy of one of the check prints, marked up and signed. If you want to demonstrate you do toolbox talks, then include a single piece of paper illustrating the register of one you did. Or if you did a lift, checked the quality of a pour or followed a programme, then take a screen shot or pdf print and include it in your package of evidence.

Figure 5.1 Examples of appendices (author's own)

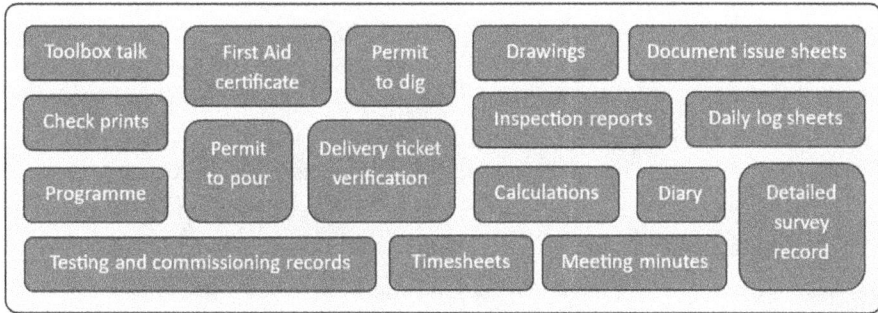

If you submit a drawing, model, safety log or add images, you must be prepared to be questioned about it. Ideally look for things with your name on, things that put you in a certain time and place, things you can point at and literally say, 'that's my name' or 'I did that.'

Once you get going on this you will find you have no shortage of evidence of how you do work and take responsibility for work done. The number of possible examples of evidence is huge. It is almost impossible to know where to start. Some of you could probably gather enough evidence from just one shift, let alone a lifetime of experience. But here are some examples.

If you are in contracting or construction

Look at how you take responsibility for a particular task. How do you manage the safety, timing and quality of the work? Do you complete daily diary sheets? Are you responsible for carrying out task briefings, checking setting out has been done or the right materials are being delivered at the right time, to the right place? How do you protect the environment or manage waste?

How are you involved and how can you prove to the two reviewers that you are a part of the construction process and that you have a key role making sure things are done correctly? Have you completed any health and safety training to ensure work is done safely?

If you are in consultancy and design

Think about how you communicate about safety through the drawings you produce. Do you communicate with clients or engineers by way of email or Microsoft Teams? How is your time managed and how do you manage it (e.g. resource charts, timesheets)?

What makes a drawing a good quality drawing? How can you demonstrate that you are involved in the quality control process? Also think about how you take office safety seriously, such as through display screen equipment assessments.

If you are working in academic research

What responsibilities do you have in the laboratories? How is the week timetabled? How do you assist students or researchers to set up experiments correctly? What measures do you need to put in place to ensure people are safe when carrying out experiments? How can you demonstrate that you are responsible for maintaining and operating standard laboratory equipment?

What procedures do you follow to keep the laboratory clean and well organised? Are you responsible for keeping detailed records of experiments, results, safety checks or any maintenance checks? Do you meet as a team on a weekly schedule? What training have you done?

If you are working as an infrastructure owner/client

What procedures or documentation do you need to complete to ensure work is planned and undertaken safely? How do you engage with different stakeholders? Are you responsible for inspecting work done? How do you track allocated budgets?

How do you record site visits and defect reports and coordinate contractors to ensure that snagging lists are addressed promptly and effectively? Are you responsible for checking that standards and specifications are followed? How do you demonstrate your role in helping to ensure that construction projects are managed safely, protecting workers and the public?

Make reference to the supporting documents when you complete the section on the Attributes. Think about how the reviewers will look at the document on screen or printed out on A4/A3 paper. You could use hyperlinks to improve how the reviewers can navigate the application if reading on a tablet or laptop. Do not be tempted to include hyperlinks to external webpages.

Remember that the overall size of your submission, which includes the form, evidence of qualifications, CPD records and appendices, should be submitted as a single pdf of no more than 10MB.

Continuing professional development (CPD)

Finally, we come to the subject of continuing professional development. In practice, most people in the industry undertake professional development daily. It is something that underpins our industry. The challenge with CPD is rarely about doing it. It is more often to do with recording it.

Did you know?

CPD is made up of a combination of two documents:

- development action plan (DAP)
- personal development record (PDR).

CPD is defined by the Institution as the 'systematic maintenance, improvement and broadening of knowledge and skills, and the development of personal qualities, necessary for the execution of professional and technical duties throughout your working life' (ICE, 2024a). The Institution's rules state that CPD is mandatory for all members.

Rule 5 of *ICE Code of Professional Conduct* states: 'All members shall develop their professional knowledge, skills and competence on a continuing basis and shall give all reasonable assistance to further the education, training and continuing professional development of others' (ICE, 2017). The ICE's guidance makes it clear that all members (not just those undergoing a review) are responsible for their own CPD; this is not something that you can pass on to your employer.

> "Get on top of your CPD straight away and make good habits in recording it, sit as many mock reviews as you can to build your confidence and ask lots of questions. If, like me, you are neurodiverse, don't be afraid to say so because reasonable adjustments can be made to support you, your style of learning and to help build your confidence."
>
> **Hannah Shewan-Friend Eng Tech FIHE MCIHT MICE**

As a candidate for the TPR you need to be aware of the CPD requirements that apply to you. The Institution normally requires you to submit three years of records, if experience allows. Although one year is acceptable, there is a warning in the ICE review guidance: 'If you do not have this, you must provide an explanation in your submission, and be able if asked by your reviewers, to show how you have maintained your competence' (ICE, 2023c). This leaves it all to chance on the day and is unlikely to produce a high level of success.

It's better to take a little time before you apply and build up a suitable quantity of evidence and submit sufficient records to demonstrate you have undertaken at least 30 hours of effective learning per year for the three years prior to your application. Your reviewers will read your CPD and may pick up on particular points or records, so ensure you can back them up.

Once qualified, there is no minimum commitment, but the ICE acknowledges that 'members generally tend to complete an average of 30 CPD hours a year' (ICE, 2024a). Whatever you think about CPD, the important thing to keep in mind is that you shouldn't ignore it. It is a mandatory requirement for technicians registered with the Engineering Council. By maintaining a personal commitment to CPD, this means employers can be reassured that their employees are developing and enhancing their abilities. It is beneficial to organisations of all sizes to demonstrate their commitment to enhancing the ability of employees to remain competitive, meet customer expectations and demonstrate a commitment to continually improve quality.

Did you know?

There are templates available for the DAP and PDR in Appendices A and B of the ICE's *Continuing Professional Development Guidance* (ICE, 2024a). The ICE strongly recommends that you adopt the templates. The DAP and PDR do not count towards the appendix page limit.

Development action plan (DAP)

This is a forward-looking document. In the DAP you should be looking at identifying your goals. You are more likely to achieve your goals if you write them down. A lot of people recommend the use of SMART targets: when setting your goals they should be specific, measurable, achievable, realistic and timely. While in principle this is a good idea, keep in mind that it's not always possible to define specific objectives and so it is ok to put in some nonspecific, open-ended goals too.

Another way of looking at this is to consider your goals in two ways. The first way is to think about what skills or knowledge you are likely to pick up or keep up in the coming year if you wish to maintain your abilities or simply continue carrying on doing a good job. Perhaps you could target some of those as goals. This reflects how you intend to maintain your competence. For example, do you need to undertake a refresher in your personal track safety (PTS) or first aid qualifications? Or will you keep up your knowledge of scaffold design standards or BIM procedures by simply carrying out work in those areas throughout the year?

The second way to reflect on your learning needs is to look further into the future and to try to imagine where you want to be or what you want to be doing in three or five years. These goals will, quite naturally, be more open-ended. It could be that you are looking for promotion to become a general foreman or a senior CAD technician. Then consider what you will need to do in the next 12 months to start to extend your knowledge and abilities and record one (or two) of these goals in your annual DAP.

However, if you have not been following a training plan focusing on the ICE, you may not have used the DAP to historically set out your development goals. In this situation it is recommended you consult your company annual appraisals or similar performance review records. It's quite possible there will be some records to demonstrate how you planned your learning and development which you can reformat into the ICE style.

To avoid any unnecessary delay, make sure you submit a current, valid DAP to the ICE. This means the deadlines for your goals should extend beyond the review day.

Professional development record (PDR)

The PDR is a detailed record of learning activities and knowledge gained. The requirement for a PDR can initially cause some alarm. But it needn't. This is something that affects technicians who are applying directly for TPR more than engineering apprentices. Apprentices often prepare for their end point assessment (EPA) after a long period of training. While under training they will have had to produce an annual record of their training achievements and ultimately had this recognised in their training review (known as Gateway) before applying for their EPA.

However, apart from civil engineering technician apprentices, technicians do not ordinarily join formal company training schemes. Therefore, it can feel more difficult to build a record of your professional development. However, employers often have in place records of learning and development. For example, most companies will put new employees through an induction process to explain the workplace and what is expected of them. This is, by definition, learning and development.

To get started, get permission to copy the company record of your training and reformat it in the style of the ICE PDR. The PDR is not simply a list of courses attended and certificates received, so you may need to spend some time reflecting on the 'key learning' and 'benefits', but this is a great place to get started.

The ICE provides comprehensive guidance on CPD (ICE, 2024a). This guidance covers various methods beyond traditional courses and certificates. It includes structured learning (such as attending workshops), self-directed learning (reading or watching recordings), informal learning (participating in webinars), on-the-job learning (gaining experience through hands-on work projects) and peer learning (engaging with colleagues). It is good to record a diverse range of learning opportunities. If you have some tangible evidence, then this can be used as proof for an audit if ever needed.

Did you know?

A key feature of the PDR is the 'effective' learning time. This is a harder question to answer than it appears. It is recommended to consider splitting a day in to chunks of time – for example, where a lunchtime talk is comparable to a one-hour period, a half day is more aligned with three hours and therefore a whole day would be six hours of effective training.

The PDR should ordinarily contain details of 'formal training' relating to the 'Health, safety and welfare' Attribute. The reference to formal training suggests that self-study or casual attendance at generic events may not be enough to satisfy this requirement.

In effect, everything you know now but you didn't know when you started working in this industry could count. The aim is not to record all the training you have received but to try to capture some of the most relevant. Education can be obtained in the classroom, but it can also be self-taught and obtained experientially. In addition to being capable of keeping yourself up to date or delivering your work better, you could look carefully at what personal qualities you are developing. Things like time management, team-working, communication techniques or learning about mental health first aid are not specifically related to civil engineering but can be incredibly valuable skills.

Career break

There are circumstances when the ICE recognises you don't need to do CPD (ICE, 2024a) – for example, when you

- are on or about to go on parental leave
- are retired and no longer involved in activity that entails the application of engineering competence
- have been unemployed for more than six months
- are suffering from any illness or disability which prevents you from undertaking CPD.

Whether it's to raise a family, travel or care for a relative, many people take some kind of break. Establishing a balance between our responsibilities at work and family commitments is often a challenge. Staying involved doesn't mean you will have the same quantity or style to your training during a break as you would have when working under normal circumstances.

Some companies, especially in relation to family-friendly policies, will have guidance on how to approach things like flexible working and how to stay in touch while on extended leave so that you can be aware of 'key updates' on things you otherwise would miss. There may also be 'keeping in touch days' so you don't feel out of the loop with work and can manage your return in a favourable way to both you and your company. These days can be recorded in your PDR to demonstrate an appropriate way of maintaining your competence.

CPD involves learning and growing even when you are not actively working. It is sharpening your skills, staying up to date and preparing for when you re-enter the workforce. Regardless of whether your career break was intended or unplanned, the knowledge and experiences you gain could have an influence on your work when you re-enter the workforce. Plus, there are plenty of ways to learn that are affordable and efficient, so you don't have to invest a lot of time or money. Webinars are a highly flexible way to access training.

Presentation preparation

For the direct application for EngTech MICE you are invited to take the opportunity to give a five-minute presentation. Although it is not necessary to submit a copy of your presentation as part of the submission, it is worth considering the presentation as part of the preparations. Good communication skills are crucial for successful completion of the TPR.

Clear, straightforward communication is the key. Think about clear and simple visual materials, as well as clear and simple spoken words. Most people put together a short PowerPoint-style presentation. Keeping the slide count down to a reasonable number will help. If you choose to do this, it is recommended to produce a handful of slides, perhaps five or six would do. A good guide to stick to would be approximately one slide per minute plus one for the introduction. With that in mind, you could be producing just five or six slides in total.

The preparation and delivery of the presentation is covered in more detail in chapter 6.

> **Did you know?**
>
> The ICE recommends using Microsoft PowerPoint; this is the most common software used. You are free to use whatever you are familiar with, provided the reviewers can see your presentation. Keep in mind, especially if presenting online, that what you may see on your screen as an A4- or A3-sized image, the reviewers may see as an A5 image if they are looking at it on a laptop. So do not clutter the screen up with too much detail or text.

Submitting your application

Before you upload your application, you should get it checked by your lead sponsor. You must pay the application fee. The application is submitted by way of the professional reviews upload portal. You should check you have completed all sections of the application form, including identifying any individual requirements. Before you upload the full application, make sure you are submitting everything that is required. A full list of the documents required is listed in *Technician Professional Review Guidance* (ICE, 2023c). All the documents should be submitted in the order listed as a single pdf of no more than 10MB.

As part of the checks make sure your two sponsors have uploaded their questionnaires. This should have been done the week before.

In summary, your application should include

- a completed application form
- a recent photograph of yourself
- evidence of your academic qualifications, if applicable
- appendices to illustrate your work
- continuing professional development (CPD) records (DAP and PDR)
- evidence of any individual requirements, if applicable.

Receipt of application

The ICE will assess your application, ensure that all necessary documents are in order, verify the completeness and assess its eligibility. If it is deemed valid, they will acknowledge receipt within ten working days. Your application will then be passed to the reviewers.

You will be told the details of the review at least three weeks before the review date. The notification will give you the names of your reviewers and the timings for the day. If your review is going to be online you will receive a formal meeting request from an ICE staff member. The meeting request will contain a Microsoft Teams link for your interview. If you do not receive a meeting request, you should contact the professional reviews team directly.

REFERENCES

ICE (2017) *ICE Code of Professional Conduct*. ICE, London, UK.
ICE (2023a) *Sponsor's Statement of Support, Version 6, Revision 5*. ICE, London, UK.

ICE (2023b) *Unspent Convictions Form, Version 3, Revision 2*. ICE, London, UK.

ICE (2023c) *Technician Professional Review Guidance, Version 3, Revision 7*. ICE, London, UK.

ICE (2024a) *Continuing Professional Development Guidance, Version 2, Revision 9*. ICE, London, UK.

ICE (2024b) *Technician Professional Review Application, Version 3, Revision 6*. ICE, London, UK.

ICE (Kes) (Kent and East Sussex Branch) (2023) *Calling All Technicians a Guide for Civil Engineering Technicians*, 2nd edn. ICE, London, UK.

Waterhouse P (2022) *Successful Professional Reviews for Civil Engineers*, 5th edn. ICE, London, UK.

Malcolm Peake
ISBN 978-1-83549-943-6
https://doi.org/10.1108/978-1-83549-940-520251006

Chapter 6
The review day

Introduction
The ability to clearly and effectively communicate your skills and experiences is crucial to the review. Nervousness, lack of clarity or inability to answer questions confidently can affect the outcome.

The review is where you show you have sufficient knowledge and experience to be a professional. It is where you show you can use proven techniques and procedures to solve practical engineering problems. Professional registration is a recognised standard that demonstrates engineering abilities, gained through knowledge and on-the-job experience.

To be successful in the review, you must satisfy all seven Attributes. This places the emphasis firmly in line with the Engineering Council which says, 'An Engineering Technician will be able to demonstrate their competence in all of the areas listed, but the depth and extent of their experience and competence will vary with the context, nature and requirements of their role' (EC, 2020).

Although this chapter is primarily focused on how to approach and think about the delivery of the presentation and the interview on the day of the technician professional review (TPR), it will be very useful for people applying for the end point assessment (EPA) too.

Did you know?

To be able to communicate clearly in English is one of the requirements to pass the EngTech with the Engineering Council and become a member of the ICE. All assessments and reviews for Engineering Council registration will be conducted in English, subject to the provisions of the Welsh Language Act 1993.

British Sign Language is an official language of England, Wales and Scotland, and is the first or preferred language of over 87 000 people in the UK. If you are deaf it is possible to undertake the interview in British Sign Language (BSL).

Online or in person?
The current application form for TPR offers a choice of an online or in-person review. This is in part because of an ongoing effort, by the ICE, to promote sustainable practices and reduce carbon emissions associated with professional reviews. Virtual interviews have been shown to reduce carbon emissions significantly by simply reducing the need to travel.

Consider carefully factors such as convenience, geography and cost as these can play a significant role in which option you choose. A substantial proportion of ordinary everyday meetings take place remotely and it is only natural that the same option is given for professional interviews.

This chapter will highlight some of the key differences to think about when considering an online or in-person review. If you feel more at ease online, go for an online review. You will be more relaxed for the interview. You will be able to join the review from the comfort of your own home or in a familiar environment such as your workplace. However, if you find it difficult to communicate online or simply prefer to meet people in person then choose the in-person option. By meeting your reviewers in person, you may find it much easier to develop a connection and as a result you will feel more relaxed.

Preparing for the review

Being able to talk about yourself and show two people you have never met before that you are a rounded professional can be tricky. Start by thinking about what is likely to happen. What do you know? The basics are described in the ICE guidance (ICE, 2023b). When preparing for the review, the first thing you should consider is what is within your control, for example:

- application
- presentation
- mock review.

To gain confidence and refine your presentation and interview skills, practice is key. If you are well prepared, then the review should not be an ordeal. It will, no doubt, be stressful, but with good practice, planning and preparation, you will be able to talk about the knowledge and experience you have built up and how you became a professional by working in the industry.

Engage with colleagues, mentors and sponsors or join preparation sessions organised by the ICE. Rehearse your responses to anticipated questions and seek feedback on your performance. This will help you identify areas of improvement and ensure that you present yourself confidently to the best of your abilities during the actual review. In addition, consider arranging a mock review.

Preparing for a TPR with the ICE requires careful planning, self-assessment and an understanding of the Institution's standards and expectations.

Other than a lack of ability to satisfy the Attributes, the most common reason for failure is due to a lack of preparation, leading to applicants feeling overwhelmed by the pressure of the day. Often reviewers record the applicant has 'failed to demonstrate' their skills. The work you put in to preparing the application and choosing suitable evidence will be key to laying the foundations for your interview.

Mock reviews

When putting your submission together your mentors/sponsors should provide you with assistance to help you prepare a suitable application and presentation, if you plan to deliver one. After you have prepared your application, make sure you set time aside for a mock review. This is an excellent way to rehearse your actual review and presentation, if you plan to deliver one. It is recommended you practise your presentation with your sponsors before your mock review.

Ideally, the mock review should be conducted by people who aren't your sponsors but have similar backgrounds. They should be trained reviewers. The mock review should be undertaken in a formal setting and follow the review process. A mock review will improve your understanding of what to expect on the day.

It will provide you with an insight into the process, reduce your anxiety and help you feel more prepared. It should help pinpoint areas where you need improvement. You will improve your time management skills. By receiving constructive feedback, you will be able to make adjustments, enabling you to present yourself confidently to the best of your abilities on the day of the actual review.

The technician professional review day

On the day, your reviewers will introduce themselves and the lead reviewer will set the scene for your review. They will explain how the review is carried out, including advice such as that they are here to engage with you, to find out what you know, not what you don't. They will often remind candidates they are not trying to catch you out or ask questions that you cannot answer.

There is an optional five-minute presentation which all candidates are encouraged to do. If nothing else, this presentation will be a great way to get the conversation going. If you have chosen to do one, then the lead reviewer will remind you about the length of the five-minute presentation and the importance of not going over the allotted time.

> "One thing I always say to the candidates on the day of the review is, 'Today belongs to you, we are not here to trip you up, just to understand where your strengths lie'."
>
> **Aaron Passfield EngTech MICE (TPR Reviewer)**

Online reviews

If you are going to be undertaking your TPR online, then it would be wise to put some time aside to find a quiet, well-lit room which is free from clutter. Make sure, for example, there is no washing in the background. Although this is not catastrophic, it can be distracting and give the reviewers the wrong impression about how professional you are during your normal working day.

When choosing your location for your review you should also make sure you will be in a place where you have a stable internet connection. If this is at work, make sure you will not be disturbed for the duration of the review; you may need to put a sign on the door.

Did you know?

If you are being interviewed online you must demonstrate you are in a room on your own, remove any backgrounds, and you will not be allowed to blur your camera.

Most people will routinely use Microsoft Teams as part of their normal day and therefore will be highly familiar with how it works. But this is not the same for everyone. If you are not familiar with it and you want to do the review online, then the ICE requires (ICE, 2023a) you to have the following:

- laptop or desktop computer
- headset or microphone and speaker
- webcam
- internet connection.

You will need to get hold of the correct software and you should make sure it works on the machine you are going to use. It would be good practice to test your equipment (camera, microphone and speakers) in advance of the day. This reduces the risk of being distracted by last-minute audio problems when your thoughts would be better focused on the review itself. Bear in mind that Microsoft Windows issues regular updates to the system which can affect the quality and stability of your MS Teams link, so check for Windows updates the day before your review.

On the day of the review, you should get yourself set up well in advance of the review. You should join the meeting, by way of the link in the formal meeting request you were sent, at least ten minutes before the start of your review. You will be admitted into a lobby and you should wait for a member of the ICE staff to invite you into the main session.

Make sure you have your passport or driving licence within easy reach so you can respond to the preliminary checks. It will not look good if you suddenly have to dash off and find your photo ID. You will be required to identify yourself, confirm you are alone and that your phone is switched off.

One way to familiarise yourself with the online platform would be to conduct an online mock interview with your mentor or a colleague. This is an excellent way to prepare and it will help identify any niggles and refine your presentation skills.

In-person reviews

If you are undertaking an in-person review, then you will find the whole process quite different. You will need to think about where your review is going to take place and how you will choose to travel there. Some people prefer to travel on the day to the review centre; others prefer to stay overnight in a local hotel. If you choose to stay overnight then choose a hotel as close as possible to the venue to reduce any concerns about arriving late.

On arrival at the venue, it is quite likely you will find the building is busy as there will be several reviews taking place on the same day and there will be other unrelated business users in the same building. The first thing to do is find the reception desk and book yourself in. This is where you will need to identify yourself using either a passport, driving licence or alternative photo ID. After this, leave anything you do not need in the cloakroom and make your way to the waiting area.

While you are in the waiting area, use the extra minutes to calm your nerves. Nerves are a natural reaction. The adrenaline release can trigger muscle tension and increased heart rate. The fear of being judged is a significant contributor to nervousness and this can cause temporary brain freeze. Forgetting what you were going to say is normal, so set yourself reasonable expectations, avoid perfectionism and practise relaxing.

There will be members of staff and a couple of senior members of the ICE in the area. They might engage in conversation with you, but rest assured, their intentions are purely to help you feel at ease. Everybody, including the staff and the reviewers, genuinely want you to be at your best and they will provide any necessary assistance. You will be in good hands.

One of the reviewers will come out to the waiting area to get you. They will call out your name and invite you to join them. From the first moment of contact you will be developing a relationship with the reviewer. Although you may not feel like it, smile.

During the short walk to the interview table your reviewer will talk informally. They may be interested in your journey and may tell you a little about themselves. They will talk about anything to try to get you to relax, so talk to them. This is not the review; be assured it has not started yet.

Your reviewer will guide you to your table. This will be one of several arranged around the room. Your two reviewers will sit opposite you at an average-sized table. Ordinarily, the reviewers sit facing into the room and you will face the reviewers with a wall behind them, possibly a window. If the latter, you must concentrate on the reviewers and not gaze out of the window!

On the table there will be some paper, a pencil and some water. Being nervous can lead to dryness in the mouth and staying hydrated can help alleviate this symptom. If you have not done so while waiting, it may be well worth taking a sip of water as you take your seat.

In summary, most reviewers are happy to interview either in person or online. The choice is yours. You should not get unduly worried about the decision. Ultimately, the TPR review is all about the Attributes and these are the same for both online and in-person reviews.

The presentation

Just thinking about the presentation will, for most people, create an unsettling feeling. It is possible, with practice, to reduce the size of this fear. It may even be possible to make the fear disappear entirely. Also, communication is different for different people. For example, some individuals with autism spectrum disorder often find it difficult to look others in the eyes, as they find eye contact uncomfortable or stressful. You must tell the ICE in your application form if such restrictions apply; they will take this into account for your review.

If you have indicated in your application that you intend to give a presentation, you should be prepared to expand on an aspect of your experience taken from the information you have already provided in relation to the Attributes.

It is highly unlikely you will be able to say everything you want to say about everything you can do in five minutes. The simple advice is not to try to do that. Clearly, if the ICE wanted you to tell them everything about yourself, they would give you a whole hour to tell your story. So, if you are not expected to tell your life story, what do you talk about?

The five-minute presentation is an opportunity for you to take control of your nerves and take a moment or two to introduce yourself to the two reviewers before they get a chance to start asking questions. During the presentation, the reviewers will not say anything. If you go over your five minutes they will stop your presentation so they can start the interview.

Whether you conduct your TPR online or in person, it is recommended you dress professionally. Choose clothing that you would normally expect someone to wear for a formal business meeting.

Did you know?

The average person speaks at about 125 to 150 words per minute, which means the average five-minute presentation is between 600 and 800 words.

Depending on what type of review you choose, the approach to the preparation of the presentation will be different. When preparing for your presentation, in addition to the content, also pay some consideration to how and where you will be presenting. Communication is not just verbal; it is also visual. Posture is important. It is a big part of our overall body language. Some people may prefer to stand when they give a presentation; others may prefer to sit at a desk. Either way, keep your back straight and shoulders relaxed. This can help you look confident. Remember: you want the reviewers to get to know you at your best.

When choosing the topic or project to talk about, keep in mind that you are trying to introduce yourself to the reviewers so they can get to know you and what you can do. Therefore, the best way to do this is to pick some work you have recently done. It may relate to the work you have highlighted in the seven Attributes. Do not forget that if you take longer than five minutes, your reviewers will stop you so that the interview can proceed.

All presentations should have a beginning, middle and end.

Think about how you will start. The reviewers already know who you are; it is not likely to be necessary to introduce yourself, but saying 'Good morning, my name is…' would not go amiss. Be bold and get straight to the point. This is simply a presentation across a table (or through a screen) to two people who will have already read your application and already know something about you.

You could take the first minute to introduce the project you want to talk about and the part of the project you were working on. Take the next minute to introduce your direct team and what your role is in that team.

Then you could spend a couple of minutes explaining what you were doing. For example, as a contractor, explain how you helped build a road or install a drainage run. As a consultant, you could describe how you developed complex CAD/BIM models to communicate information using 2D or 3D models. If working as an academic, you could explain how you ran a laboratory, or if working as a client, how you took overall responsibility on a project.

When preparing your presentation, you need to remember you are telling the reviewers about you, your roles and responsibilities and how you ensure the 'job is done right first time'.

After that you could look at wrapping up the presentation. For the last minute you could have a think about some of the practicalities surrounding your work. This may be some of the more routine things that need to be done: things like ensuring safety briefings are undertaken or CAT and Genny surveys recorded. You could evaluate the project or your role, or something that went right or wrong. Maybe you could talk about hold points, inspections you do, or need others to do, permits to load or the check print process.

This type of information lets the reviewers know you are a dependable person. It lets them know you can get work done properly and take quality checks into account or that you can work safely either on your own or as part of a team. This could be as simple as making sure you don't just wear PPE, but you wear the right PPE for the job, or, depending on your responsibilities, making sure that your team is wearing the right PPE.

Everybody's personality is different and how you choose to portray yourself or get your message across will be different from other people. If you are not familiar with using PowerPoint, then do not feel obliged to use it. Use paper or simply just talk face to face or to the camera. If you need a few notes to make sure you keep on track, then prepare a short list of the key things you want to include.

I'm sure you will have heard the phrase 'a picture says a thousand words'. Keep things simple. Don't be tempted to clutter up the slides with loads of information. Look to have large images and a handful of words to help the reviewers visualise what you are saying.

Lastly, have a think about how you want to finish the talk. Do think of how you hand over to the reviewers. Perhaps you could simply round things off by saying 'I hope you found that interesting, I'm happy to answer any questions.'

Online presentations

When faced with the prospect of delivering a presentation on screen, all our perceived flaws could feel like they are magnified by the camera and, for some, fear can creep in. On the day itself, the time will go very quickly and there is pressure to start and stop your presentation in the time allowed. Delivering the five-minute presentation can be both exhilarating and nerve-racking. This can feel like quite an intense and unusual situation.

"Plain and simple, EngTech MICE is formal recognition of your competence and contribution to the profession."

Tim Brownbridge EngTech FICE

At this stage, you will have prepared a good script and prepared the visual slides. The only thing left to do is to think about how the live performance will go on the day. Keep in mind your reviewers' perspective and how they will be engaging with you.

When presenting to your reviewers, remember to look them in the eyes. This is easier said than done when presenting online and even harder if using more than one screen. Make sure you are facing the camera and try to relax. When online, you are acting as a presenter and therefore it is best to try to look down the lens of the camera (ICE, 2023a). This way, it is clear you are addressing the two reviewers. However, you may find it more comfortable to look a centimetre or two above your camera. Also do not forget to smile. A smile is a great way to engage with your reviewers. Try to look as though you are pleased to be there.

In-person presentations

The same rules of engagement should be considered for the presentation (if you have chosen to do one) as outlined in the online section. You are encouraged to use visual aids and you should consider the scale of your presentation. Anything you use to illustrate your work should be of an appropriate scale. Keep in mind you are speaking to two people across a table and therefore a practical consideration would be to use either A3- or A4-sized documents.

There can be a temptation to do the presentation on a laptop. This is permitted; if you do, you will need to make sure your laptop is fully powered. The ICE cannot guarantee there will be access to an external power supply.

"On the day I was understandably nervous. I gave myself the best chance of success by preparing all my notes and presentation well in advance and revised them several times beforehand. Once I started the interview it went well, and the interviewers were very reasonable and easy to talk to."

Rob Ehren EngTech MICE

In summary, if you choose to do a presentation online or in person, all you are doing is talking about yourself and your day job! Sometimes, a little pressure can help to focus the mind and enable you to concentrate on what the key message is that you are trying to deliver.

If you take time to focus on the purpose of the five-minute presentation and perhaps consider it more like a five-minute introduction, then you may find it easier to approach and find it beneficial to you (and the reviewers) in starting the discussion. So, try not to be nervous: keep calm, breathe easy and take the opportunity to introduce yourself to the reviewers so they know something about what you consider to be a good day's work and a job well done.

The interview

The interview will be a discussion with your reviewers based on both the content of your presentation, if you give one, and the information you provided in your application. If you have approved, formal academic qualifications, the whole review will usually last up to 45 minutes.

Did you know?

If you have been given an extended interview (up to 60 minutes) because you do not have the necessary formal qualifications, there will be a greater focus on the first Attribute: understanding and practical application of engineering.

Essentially, the reviewers are looking to determine whether you have gained an equivalent education in the world of work compared to an applicant with a more formal, approved academic qualification. You should take the opportunity to demonstrate your knowledge and understanding of engineering principles in your presentation and interview. Otherwise, all aspects of the application and review are the same.

If you choose to give a presentation, your reviewers will engage in a conversation that draws on specific aspects of it. They will then shift their focus to your application. If you do not give a presentation, they may ask you to introduce yourself and then proceed with questioning the content of your application. Before the day you should practise delivering clear and concise explanations of what you do. Think about your work and what you want the reviewers to know about what you do.

There are seven Attributes and people often ask which are the most important ones. What they are asking is which Attributes they need to pass. The fact is that all Attributes are equal and if you fail any, you fail the whole review. While correct, this is often unhelpful. What people really want to know is what the reviewers are really looking for. The reviewers are not trying to catch you off guard; rather, they aim to have a professional conversation where you can demonstrate your skills and expertise.

> "It's worth doing a little research and gaining some knowledge of the institution you want to be a member of. For example, before your big day think about why you want to be a member, read up on the six rules of conduct and find out who is the current president!"
>
> **Aaron Passfield EngTech MICE (TPR Reviewer)**

The reviewers will not necessarily ask you questions in the order of the Attributes. Some of the questions will be subtle and others more obvious. The reviewers are checking you are who you say you are and will pose questions to check the details behind the work you describe.

It is important not to waffle; the temptation can be to keep talking until you are asked another question. This can trip you up or lead you to talk about aspects beyond the question that was posed. Try to keep your answers concise and focused on answering the specific question. Your reviewers only have a relatively short period of time to satisfy seven Attributes, so don't worry if they cut you off or move you along.

Towards the end of the interview, you will get a sense that the process is drawing to a close. However, the reviewers are likely to ask you if you have any questions or whether there is anything else you would like to add. This is another chance for you to engage with the reviewers on a professional level, so don't miss out on the opportunity.

This may come as a surprise to you, but the interview can often be a positive experience. Reviewers enjoy meeting good candidates. It can be a real pleasure meeting people who are practical problem-solvers, capable of undertaking their role in a confident and competent manner.

There are times when the process can be less pleasant; reviewers do not enjoy meeting people who have not prepared or are not up to standard. You would probably not enjoy that sort of interview either, so prepare and proceed with caution. Remember: your sponsors have full confidence in your capabilities and competence, so you ought to take heart from this and believe in yourself.

> "The hardest thing to do is to sell yourself. You need the confidence to say this is what I do, it matters, and I should have the professional recognition for it."
>
> **Simon Dunbar EngTech FICE Tech IOSH MPWI**

Observers

It is worth noting that there may be an observer at your review. This can happen in both online and in-person reviews. If there is a fourth person attending, you will be notified on the day.

The observer is present solely in an observational capacity and does not actively participate in the review process. They are not there to review you; they could be an observer auditing the process from the ICE or the Engineering Council.

If online, you will notice an additional face on the screen. At an in-person review, the observer will sit just within your field of vision at one end of the table. They will not participate in the review.

Summary

This is your opportunity to demonstrate your ability to do the job you do. Overall, you need to demonstrate you have an appropriate balance of knowledge, experience and abilities that you use when applying proven techniques to perform your role.

Table 6.1 summarises the similarities, differences and details that affect the two different types of TPR review day.

The questioning that follows your presentation, assuming you chose to do one, is not intended to be an interrogation. The reviewers are not trying to catch you out or trip you up. They will start from the assumption that you are competent, and they will want to find out more to confirm that opinion.

The reviewers will ask open questions and try to gain an understanding of what is behind the facts. They will want to get to the bottom of why you do what you do, what the knowledge that underpins it is and where you have applied your knowledge. This will encompass the full range of Attributes and show how each of these areas impact on each other and your ordinary working day. You will be able to show that you are not just technically able to do the job but you can also manage your time, follow programmes, follow quality controls, work in teams and take responsibility for your safety and your impact on the environment. In short, you will

Table 6.1 Summary of details of the two different review days

	TPR	TPR (missing academic base)
45 minutes	5-minute presentation (optional)	5-minute presentation (optional)
	Up to 45-minute interview on whole application to confirm you meet all the Attributes	Up to 60-minute interview on whole application to confirm you meet all the Attributes, with a focus on understanding and practical application of engineering
15 minutes		

demonstrate that you are a rounded professional. The reviewers' focus will be to confirm that you meet all of the Attributes at EngTech level. To be successful, both reviewers need to be satisfied you have demonstrated all the Attributes.

The interview is your time to shine. Visualise the day, practise, aim for a good posture and seek to hold eye contact with the reviewers. The better prepared you are, the more relaxed you will be on the day. The more relaxed you are, the more likely your reviewers will be able to get to know you. If you know your story, you will be ready to share better.

REFERENCES

EC (Engineering Council) (2020) *The UK Standard for the Professional Engineering Competence and Commitment (UK-SPEC)*, 4th edn. EC, London, UK.

ICE (2023a) *Online Professional Review: Candidate Guidance, Version 2, Revision 1*. ICE, London, UK.

ICE (2023b) *Technician Professional Review Guidance, Version 3, Revision 7*. ICE, London, UK.

Successful Professional Reviews for Civil Engineering Technicians

Malcolm Peake
ISBN 978-1-83549-943-6
https://doi.org/10.1108/978-1-83549-940-520251007

Chapter 7
Apprenticeships and the end point assessment

Introduction to apprenticeships

A civil engineering apprenticeship is a programme that combines practical, on-the-job training with academic learning. As an apprentice you will be working under the guidance of experienced technicians and engineers, gaining hands-on experience of the industry.

An apprenticeship begins with an induction, where an apprentice is introduced to their programme. This is followed by a structured training period where they develop the necessary knowledge, skills and behaviours (KSBs) through practical experience and coursework. Once the training is complete, the apprentice reaches the Gateway stage, where their progress is reviewed by the employer and training provider to ensure they are ready for the end point assessment (EPA). The process concludes with the EPA, which evaluates the apprentice's overall competence and readiness to independently perform their role.

> **Did you know?**
>
> The Institute for Apprenticeships and Technical Education (IfATE) recommends the typical duration from start to Gateway is 36 months. This does not include the time spent preparing the application and sitting the end point assessment (IfATE, 2021b, 2022, 2024a, 2024b).

This chapter will focus primarily on the period after Gateway through to the EPA. We will look at the EPA application process, the portfolio of evidence, the technical project brief and finally there will be some advice about what to do on the day of your EPA. There are two current apprenticeships:

- Civil Engineering Technician (CET) level 3, version 1.1 (ST0091)
- Civil Engineering Senior Technician (CEST) level 4, version 1.1 (ST0046).

These are more commonly known as either a level 3 or level 4 apprenticeship and that is how they will be referred to in this chapter. This chapter will provide an overview of both apprenticeships and, where appropriate, separate specific guidance on the level 3 EPA and the level 4 EPA.

In combination with the EPA, it is possible to be awarded EngTech MICE. The Institution of Civil Engineers is the only end point assessment organisation able to offer end point assessments for these two apprenticeships and award professional recognition as EngTech MICE. There is very little difference between an EPA on its own and an EPA with EngTech MICE included.

This chapter focuses on apprenticeships associated with contractors, consultants and clients.

The book is for technicians so there will be no focus on the level 6 civil engineer degree apprenticeship.

At the time of writing this book, the level 3 version 1.1 is still current for apprentices who started after 14 July 2021. The version history for the level 4 apprenticeship is more complex. Table 7.1 best illustrates the dates associated with the different versions.

Guidelines for the level 4 version 1.2 and 1.3 EPA have yet to be issued but as the knowledge, skills and behaviours (KSBs) have remained consistent across versions 1.1, 1.2 and 1.3, it is anticipated the EPA will also remain consistent.

The updates have mainly been to ensure the apprenticeship programme runs smoothly, remains relevant and prepares people for working in the civil engineering industry. Some of the changes are outlined below.

- There is a need to undertake a comprehensive review of progress and readiness when passing through Gateway. This is to ensure you will be ready for the EPA.
- The overall application process is being streamlined to improve efficiencies. This includes clearer deadlines and assessment dates.
- There will be more emphasis on the inclusion of sustainability and environmental impacts when responding to the technical project brief.

"The ICE process teaches you useful, practical life skills such as presentation, interview and Q&A skills. Granted, the process was somewhat difficult and doing this at the age of 20 didn't make it any easier, but, if you power through it and really put the effort in, the result is extremely rewarding.

It gave me the confidence boost I needed to pursue bigger and better things. In reality, it's more than just a title. For me it is the first step to help me realise my dream to become a chartered civil engineer."

Ibrahim Kapasi EngTech MICE

Table 7.1 Version history for the level 4 apprenticeship (IfATE, 2024b)

Level 4 version	Earliest start date	Latest start date
1.3	05/09/2024	Not set
1.2	03/04/2024	04/09/2024
1.1	01/07/2022	02/04/2024
1.0	27/03/2018	30/06/2022

Flow map

The flow map in Figure 7.1 illustrates the whole apprenticeship process, in one page, from the very beginning with registration through Gateway and on to the final end point assessment.

Initial training period

Typically, the apprentice training programmes are anticipated to last up to 36 months. The first two years are linked to either a level 3 or level 4 academic qualification (e.g. BTEC or HNC). This is then followed by a period of ongoing training and development intended to consolidate your knowledge, refine your skills and address any gaps where you need more practice or knowledge. Throughout the whole apprenticeship you will record both 'off-the-job' and 'on-the-job' training.

The overall purpose is to help you improve your competence and gain confidence in your abilities before you leave the training programme and start preparations for your EPA. It's worth keeping in mind that the skills assessed in an EPA are very closely linked to the skills assessed in a technician professional review.

Training 'off-the-job'

An apprenticeship is defined as a job with training. The term 'off-the-job' training is used in legislation. To be eligible for government funding, apprentices must spend at least 20% of their normal working hours (capped at 30 hours per week for funding purposes only) over the planned duration of the apprenticeship practical period on 'off-the-job' training. This means that a *minimum* requirement for apprentices working 30 hours or more a week is an average of 6 hours per week.

In the UK, most workers who work full-time are legally entitled to 28 days' paid annual leave a year. This is equivalent to 5.6 weeks of holiday. Usually this includes 20 days of leave plus 8 bank holidays. Assuming on average you complete 1 day a week 'off-the-job' training and since there are 52 weeks in a year, roughly calculated you should be recording 46.4 days or 278 hours of training a year (DfE, 2024).

You should regularly record your 'off-the-job' training during the initial training period. The majority of this will be related to the teaching at your college work. In addition to lectures, you can include revision and time spent writing assignments. You will also receive training by

Figure 7.1 The whole apprenticeship process (author's own)

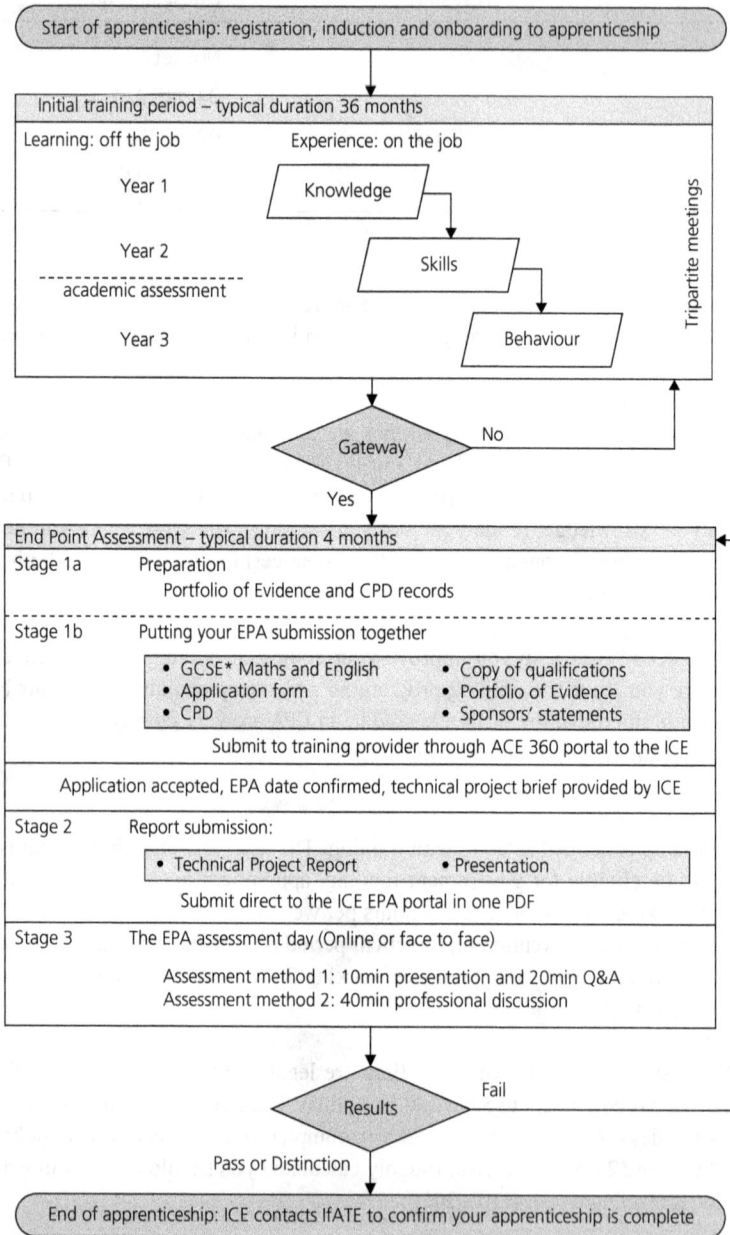

Start of apprenticeship: registration, induction and onboarding to apprenticeship

Initial training period – typical duration 36 months

Learning: off the job   Experience: on the job

Year 1   Knowledge

Year 2   Skills

academic assessment

Year 3   Behaviour

Tripartite meetings

Gateway — No

Yes

End Point Assessment – typical duration 4 months

Stage 1a   Preparation
Portfolio of Evidence and CPD records

Stage 1b   Putting your EPA submission together

- GCSE* Maths and English
- Application form
- CPD
- Copy of qualifications
- Portfolio of Evidence
- Sponsors' statements

Submit to training provider through ACE 360 portal to the ICE

Application accepted, EPA date confirmed, technical project brief provided by ICE

Stage 2   Report submission:

- Technical Project Report
- Presentation

Submit direct to the ICE EPA portal in one PDF

Stage 3   The EPA assessment day (Online or face to face)

Assessment method 1: 10min presentation and 20min Q&A
Assessment method 2: 40min professional discussion

Results — Fail

Pass or Distinction

End of apprenticeship: ICE contacts IfATE to confirm your apprenticeship is complete

*Any GCSE with English (regulated by Ofqual, CCEA and Qualifications Wales) in the title where English is the primary language, unless explicitly stated or 'English Functional Skills level 2' and any GCSE with Mathematics in the title. Minimum acceptable grade for functional skills, pass and for GCSE, grade C or 4 (from 2017, where 9 to 1 grading scale is used).

shadowing colleagues and being mentored at work. In addition to recording the hours spent 'off-the-job' you will need to record progress against the learning outcomes. These come in three stages and are commonly known as the KSBs (knowledge, skills and behaviours).

Ordinarily, you cannot include training towards a level 2 in English or Maths or anything that does not build knowledge, skills and behaviours relevant to the apprenticeship. You cannot include progress reviews, mock EPA testing or anything that takes place outside of your normal working hours. This is a complex issue, and this book is not intended to delve into the full details. If you want to know more, it is recommended you speak to your college, your company or refer to government guidance on this subject published by the Department for Education (DfE, 2022).

Did you know?

Off-the-job training includes anything you do in your *normal working hours* that teaches you *new* knowledge or a new skill that is *relevant* and contributes to the successful completion of your apprenticeship.

For example, if you were training to be a bridge inspector and you were learning to abseil during the week as an ordinary part of your working day so that you can carry out inspections of bridges, then this would count as off-the-job training. But if you were a CAD technician learning to abseil at the weekend for the adrenaline rush and sense of adventure then you cannot count it because it is outside your normal working hours and it may not be relevant to your apprenticeship.

Training 'on-the-job'

On-the-job training is training received by the apprentice for the sole purpose of enabling them to perform the work for which they have been employed. This is intended to mean training that does not specifically link to the knowledge, skills and behaviours set out in the apprenticeship (DfE, 2024).

Tripartite meetings

To help keep track and monitor progress there will be a series of tripartite meetings throughout the apprenticeship. These are held every 12 weeks, which is typically at three-month intervals. These meetings involve you, your employer and your training provider. The meetings are essential to ensure progress is reviewed and to establish a supportive, collaborative environment. As an apprentice you must record off-the-job training, your progress against the KSBs and attend tripartite meetings from the start of your apprenticeship until Gateway.

By recording what has been achieved, evaluating the current status and looking ahead, everyone is kept informed and engaged. Everyone should engage in these meetings.

Did you know?

Each person should typically contribute the following:

Apprentice

- progress: share progress made at college and at work
- feedback: provide feedback on academic and industrial training received
- goals: discuss future objectives with mentors and how they will be achieved.

Industrial mentor (provided by the employer)

- progress: provide insights into apprentice's performance, building evidence of KSBs
- support: ensure the apprentice has the necessary learning support at college
- workplace: identify future opportunities to build key skills that align with the KSBs.

Academic mentor (provided by the college/training provider)

- progress: report on academic performance and check 'off-the-job' and KSB documents
- support: identify future goals, opportunities and support from employer
- plans: guidance on future academic requirements e.g. exam dates, project deadlines.

Gateway

When you reach Gateway, this completes the practical period of your apprenticeship. Gateway is not the end of the apprenticeship; it is a crucial checkpoint. It demonstrates you have completed your initial training period and have acquired the knowledge, skills and behaviours outlined in the apprenticeship standard.

The Gateway meeting is essentially the final tripartite meeting between you, your college and your company. This is a formal process where your employer and training provider review your final set of training records. If this is all completed and agreed, you can then move on to the end point assessment (EPA). In making this decision, your employer will take advice from your training provider, but the decision must ultimately be made by your employer.

By passing through Gateway, you are indicating you are ready to move on to the final assessment where you will demonstrate your competence in your role in the EPA. The EPA interview day is ordinarily undertaken within approximately four months of Gateway.

It can be useful in the Gateway meeting to define your target date for your EPA. The college will be involved in processing your EPA application so it would be good to discuss your next steps. Having a clear, well-defined plan of action and a good appreciation of the objectives will be beneficial to you, your company and your training provider. It will enable you to consider the resources you need to carry out and manage your work in line with the EPA application process. If planned well, you will end up with a good quality application delivered on time to the right standards.

Did you know?

To pass through Gateway you will need to provide:

- evidence of English and Maths level 2 (e.g. GCSE)
- evidence you hold a BTEC level 3/HNC level 4 qualification (certificate from college)
- record of your off-the-job training
- record of your completed knowledge, skills and behaviours (KSBs).

Overall, this book is intended to provide a comprehensive guide to enable people to understand how to apply and be successful in their technician professional review (TPR). Therefore, to align with that goal, this chapter will similarly focus on applications for a successful end point assessment (EPA). It will be assumed you will have completed your period of training and already completed Gateway. The EPA is like the final exam for apprenticeships. It is where you prove you have learned what you need to know to join the industry and continue to work as a professional.

"When I completed Gateway and started my preparations for the EPA, I felt very apprehensive about the whole process. I felt like there was so much to do and not enough time to do it."

Ayomikun Ajayi EngTech MICE

The EPA

Shortly after concluding Gateway, you are likely to start preparing for your EPA. The first thing to do is make sure you are looking at the latest guidance, for example, *Civil Engineering Technician (CET) Level 3, version 1.1* or *Civil Engineering Senior Technician (CEST) Level 4, version 1.1* (ICE, 2004a). They provide detailed guidance, explain what to submit and outline the whole application process.

You also need to decide when and where you would prefer to be assessed. The assessments usually take place three times a year (February, June and October). They can be held in person or online.

Did you know?

The EPA day includes two methods of assessment:

- assessment method 1: technical project presentation followed by question-and-answer
- assessment method 2: professional discussion based on portfolio of evidence and CPD.

Putting the application together

Once you have established which guide will help you and when and where you wish to sit your EPA, the next step is to start putting your application together. In addition to the application form there are some key elements which could take some time to prepare.

Did you know?

You will need to submit:

- evidence of English and Maths level 2 (e.g. GCSE)
- certified copy of your qualification from college
- application form – identifying 'technical specialism and scenario'
- portfolio of evidence
- CPD records (DAP and PDR) (optional if applying for EngTech MICE)
- two sponsor's statements (optional if applying for EngTech MICE).

Qualifications

The first two elements of the application package should be easy to put together. Any GCSE with English (regulated by Ofqual, CCEA and Qualifications Wales) in the title, where English is the primary language, unless explicitly stated or 'English Functional Skills level 2', and any GCSE with Mathematics in the title, are suitable. Minimum acceptable grade for functional skills: pass and for GCSE: grade C or 4 (from 2017, where 9 to 1 grading scale is used).

Your college should have a copy of your BTEC or HNC qualifications. Your certificate should be signed by a professional person. Ordinarily, people ask one of their sponsors or a tutor from the college.

Did you know?

The person signing the copy of the certificate should write the following:

"I confirm this to be a true copy of this applicant's qualification".

Signature:

Print name:

Employing organisation/college:

Position:

Contact telephone number or email:

ICE membership number (if applicable):

The application form should be read and completed carefully. The other two key elements are the portfolio of evidence (PoE) and the continuing professional development records (CPD). Like the sponsors' statements, you only need the CPD records if you are applying for membership of the ICE. Although obtaining EngTech MICE is not a mandatory part of the apprenticeship, based on the fact you are reading this book it will be assumed you are interested in joining the ICE.

It can take three or four weeks to pull all this information together. If you have been building up your records of the KSBs and CPD throughout the initial training period then you may find the process much less stressful and quicker. We will now go into more detail on these three elements.

Mock end point assessments

When putting your submission together, your mentors/sponsors should provide you with assistance to help you prepare a suitable application and respond to the technical project brief. After you have applied for your EPA and later submitted your response to the technical project brief, make sure you set time aside for a mock EPA.

Ideally the mock EPA should be conducted by people who aren't familiar with your work but do have similar backgrounds. They should preferably be trained assessors with the ICE or at least familiar with the procedures. The mock EPA should be undertaken in a formal setting and follow the full EPA process. This will improve your understanding of the day.

The mock EPA should help you feel more prepared and consequently help reduce your anxiety. It should help pinpoint areas where you need improvement. By receiving constructive feedback, you will be able to make adjustments, enabling you to present yourself confidently to the best of your abilities on the day of the actual EPA.

The application form

As you look through the form, in addition to simple questions about your name, address and other personal details, you will find other questions related to employee type, technical project options, CPD records and supporting information (see Figure 7.2). There are three main sections and we will look at each one in detail.

Figure 7.2 Extract from section 1 of the application form

A. Employment type (select only **one**)

Contracting and Construction	☐	Consultancy and Design	☐
Academic Research	☐	Infrastructure Owner / Client	☐

B. Civil engineering project subject focus, area of technical specialism (select **one** scenario)

Bridges Scenario:	☐	Dams / Reservoirs Scenario:	☐
Environmental planning / engineering Scenario:	☐	Building, Structures Scenario:	☐
Airports Scenario:	☐	Water Supply / Sewerage treatment / Drainage and networks Scenario:	☐
Geology, Geotechnical and Ground Engineering, Tunnelling Scenario:	☐	Railway Systems and infrastructure Scenario:	☐
Offshore Engineering Scenario:	☐	River, Coast, Marine, Docks and Harbours Scenario:	☐
Transportation Planning Scenario:	☐	Highways & Traffic Engineering Scenario:	☐
Regeneration and Development Scenario:	☐	Energy Services Scenario:	☐

(ICE, 2024b; ICE, 2024c)

Section 1 – completed by your employer

Ordinarily, your line manager or your industrial mentor should complete this section. They should put in both your name and their own. They should also add in details of the employing organisation, the position they hold and their contact email address.

By completing this section and signing it, they are confirming that you have completed your academic studies, you are suitably experienced and you have all the relevant knowledge, skills and behaviours. They will also identify what *technical specialism* is closest to your experiences and choose a *suitable scenario* for your technical project brief. Essentially, they are confirming you have gone through Gateway.

This element of the application form should be considered carefully. The ICE provides a list of project scenarios. These change from time to time, so it is important you have the latest version. To ensure you are up to date, you should contact the ICE directly by writing to epa@ice.org.uk and ask for the most recent version (ICE, 2024f).

The ICE will use this information to prepare a *scenario* and *challenge* for you to respond to. You will then be responsible for preparing a technical project report and presentation, based on a real, work-based civil engineering challenge.

Some example scenarios and corresponding codes your employer may choose include the following.

- **Scenario code B1:** a proposed route for the construction of a new road for a local council which will cross a railway line and is adjacent to a power station with places for expansion as well as a new residential development.
- **Scenario code BS2:** the development of a four-storey scientific building for a new science park on a brownfield site which is bounded by a river, an existing light industrial building, a highway and woodland. The new laboratory building will be located adjacent to the river.
- **Scenario code R3:** optioneering for the report or replacement of a steel rail bridge across a river.

If none of the scenarios proposed by the ICE relate to the type of real-life work you do then your employer should contact the EPA team and consult with them on alternative options.

Section 2 – about you

You should find it relatively easy to complete most of this section. Many of the questions are directly related to you, such as your name, date of birth, nationality and home address. There are some slightly more complicated ones related to your employer and training provider. But these are primarily still all straightforward, and your employer and college should be able to assist you.

However, one or two of these questions will take a little more time and effort to answer. For example, you will be asked for your ICE membership number. If you have already registered as a student, then you will have a number. If you have not registered as a student, go to https://myice.ice.org.uk/register. In a matter of minutes, you will have generated a unique eight-digit personal identification number.

The unique learning number (ULN) is a ten-digit number used by the Department for Education. Most schools, colleges or training providers print the ULN on certificates. If you do not know yours, your college should be able to provide your ULN reference. Finally, you are asked for a UKPRN number. The UK provider reference number (UKPRN) is not unique to you; it is an eight-digit unique identifier assigned to your college or training provider. It always starts with 1. Your college should be able to provide this information.

If you are planning to continue your academic education after completing your EPA and progress onto another apprenticeship (e.g. level 4/level 6) or further (e.g. HNC) or higher (e.g. BSc/BEng) education, then you should indicate your plans to the ICE. By providing details of your next course, the ICE administrations team will endeavour to ensure your EPA is completed in time to allow you to progress your education. This can be especially important if you are planning on sitting an EPA in June and will be starting the next phase of your education in the following October.

Towards the end of this section, you will be asked to provide some diversity data (if willing) and details of any individual requirements. You will also need to list your academic qualifications. You must include a verified copy of your academic qualifications in civil engineering from your college.

EPA session and format – online or in person

You need to decide when and where you would prefer to be assessed. You should check out the ICE 'key dates' on the website and choose a suitable session. These usually take place three times a year (February, June and October). Please note that assessments can be held in person or online. So, you need to identify not only your preferred session but also your preference, whether online or in person.

You need to confirm you are happy for the ICE to contact the Institute for Apprenticeships and Technical Education (IfATE). Your approval is necessary for the ICE to make this request on your behalf, ensuring that all personal data and actions are authorised by you. This process helps maintain transparency and compliance with data protection regulations.

Supporting information

Almost as a reminder, the application requires you to include a copy of your level 2 in English and Maths (or equivalent), evidence of your academic qualification in civil engineering and your PoE.

The evidence of English and Maths and the college qualifications should be uploaded by the training provider. The portfolio of evidence should be in pdf or MS Word format. The file, including any appendices, should not exceed 10MB.

Sponsors' statements and CPD should be included in the package of information sent to the training provider so they can upload them using the ACE 360 platform.

Section 3 – application for EngTech MICE

If you wish to be assessed for both your apprenticeship and Engineering Technician qualification at the same time, then simply confirm you have the relevant qualifications, CPD records and sponsors. In practice, most people in the industry undertake professional development daily. It is something that underpins our industry. The challenge with CPD is rarely about doing it. It is more often to do with recording it. The ICE provides comprehensive guidance on CPD (ICE, 2024d). CPD is made up of a combination of two documents: the development action plan (DAP) and the personal development record (PDR). We will go into more detail on this later in this chapter.

Did you know?

There are templates available for the DAP and PDR in Appendices A and B of the ICE's *Continuing Professional Development Guidance* (ICE, 2024d). The ICE strongly recommends that you adopt the templates. The DAP and PDR do not count towards the appendix page limit.

If you are able to find suitable sponsors, it is well worth applying for the EngTech registration. They will need to be familiar with your work. They must both be professionally recognised and registered at either EngTech, IEng or CEng grades of membership by a professional engineering institution.

It is worth taking time to consider the sponsor's statement of support (ICE, 2024e). You are not responsible for preparing the statements of support. All you need to do is make sure you include the names and personal identification numbers of your two sponsors on the application form. All applications for membership must be supported by *two* sponsors. There is more information on sponsors later in this chapter and in chapter 3.

Portfolio of evidence

The PoE is a key component in the EPA application. The requirements for the PoE are outlined in the ICE guidance (ICE, 2024b, 2024c). The ICE has developed a template and this identifies the four main areas your portfolio must cover. These group the KSBs together, making it a little easier to present the evidence you have available to show how you are able to solve well-defined problems. You can include up to 12 pieces of evidence across the four main areas.

Did you know?

The ICE PoE template identifies four main areas your portfolio must cover.

Level 3 portfolio of evidence:

1. using technical software to present civil engineering information
2. health, safety and welfare
3. project management
4. personal and professional practice.

Level 4 portfolio of evidence:

1. design, technology and modelling in civil engineering
2. project management and safe systems of working
3. roles, responsibilities and engagement with others
4. personal and professional practice.

The PoE should be concise and focused on quality not quantity. There is no strict page limit or word count provided by the ICE. A reasonable suggestion would be to limit your response to either 200–400 words in support of each piece of evidence or 2–4 pages per area. Either approach would produce a 20–40 page report. Alternatively, you could aim to produce between 2000 and 4000 words in total. A PoE of this size should provide ample evidence for the assessors to appreciate your experiences and enable them to ask relevant questions on the day of your EPA.

Your portfolio must be mapped against the relevant KSBs. There are different criteria for level 3 and level 4 PoE; these are summarised below. It is recommended you pick one or two projects or engineering challenges and use them, if possible, throughout the whole portfolio as an example of what you can do, now you know how to behave as a trained professional.

Did you know?

The portfolio of evidence only covers some of the KSBs.

Level 3 KSBs:

Knowledge: K1 K2 K3 **K4** **K5** K6 K7 K8 **K9** K10 **K11** **K12**

Skills: S1 S2 **S3** **S4** S5 S6 **S7** S8 **S9** S10 **S11** **S12**

Behaviours: **B1** **B2** B3 **B4** **B5** **B6**

Level 4 KSBs:

Knowledge: K1 K2 K3 K4 K5 **K6** **K7** **K8** K9 **K10** **K11** K12
K13 K14 K15 **K16** **K17** **K18** **K19** **K20**

Skills: S1 S2 S3 S4 S5 **S6** S7 **S8** **S9** S10 **S11** **S12**
S13 **S14** S15 **S16** **S17**

Behaviours: **B1** B2 **B3** B4 **B5** **B6**

(IfATE, 2021b, 2022, 2024a, 2024b)

In the following pages, each of the four areas of the PoE has been looked at separately. For each area an overview has been prepared, including an interpretation, alongside some sample indicative responses linked to the different KSBs. They are real-world examples to showcase some of the typical duties that various technicians from different employment types do. By showcasing real-life examples of what others do, it is hoped it will make the application process more relevant and stimulate ideas on how you could create your own responses.

It is impossible to provide an example of every type of job, role and situation in the industry. The sample responses should be discussed with your mentor/sponsor so you can appreciate how they may relate to you and your own role.

This is intended to provide you with some inspiration to enable you to produce your own response relevant to your employment, your role, your responsibilities and your experiences.

The KSBs have also been cross-referenced to the Attributes of a technician (EngTech MICE). This is not a necessary part of the application, but it does mean you can use chapter 4 of this book to delve deeper and look at further examples.

Once complete, both you and your mentor should sign the PoE to confirm the evidence is your own work and is directly attributable to you.

Level 3 portfolio of evidence

1. Using technical software to present civil engineering information

Overview

This first element is key to showing how you use technical software to do your role. This is at the centre of your role and is how you can show your practical knowledge and experience.

You will be the link between theory and practice, working alongside engineers on well-defined civil engineering designs. This will be your opportunity to provide two or three examples of what you can do and how you do it to ensure it is done correctly, in a detailed and precise way.

EPA requirements

K4	Technical drawings, modelling and designs, using computer-based software packages, such as computer aided design (CAD) or building information modelling (BIM), and their use in the sector	1(b)
S3	Operate appropriate software packages for data gathering and analysis, such as computer aided design (CAD) or building information modelling (BIM), to create technical drawings, models and designs using relevant conventions and engineering terminology	1(a)

The KSBs have been mapped across to the Attributes as featured in the ICE publication (ICE, 2021).

Interpretation

You will need to show how you apply your understanding of CAD or BIM to your engineering. Below are some short examples you could use to expand your ideas on how you may link your work to the EPA requirements. Remember to clearly define the problems you solve and link your claims to evidence to demonstrate what you are capable of doing.

Have you been responsible for coordinating the construction phase of an element of work? Did you use BIM to manage the information or CAD to find out setting out information? Have you been preparing or modifying drawings? Have you been involved in preparing tender drawings, construction drawings or as-built records? Do you use a common data environment to prepare project plans and plan resources for consultants and contractors?

Sample responses

Contracting and construction: setting out – drainage

We use a BIM model to manage all our data. The 3D models help me to visualise the entire project and extract key data. For example, I use the drawings to find out the location of manholes and alignment of the drainage runs. I also get the specifications for the pipe sizes and material such as PCV or concrete. I then set out the invert levels and any changes in gradient.

In addition, I can use the BIM model to get information about nearby utilities. I then check this information on site and do a CAT and Genny check to accurately identify the location of underground services.

Consultancy and design: CAD technician producing drawings

I work closely with a geotechnical engineer. I was responsible for producing the construction drawings for soil nails to stabilise a railway slope. I made sure they are set out geometrically and spatially correct. My drawings include notes on the diameter, length of strands and spacing of nails.

Infrastructure owner/client: asset inspector

I am responsible for carrying out inspection work along the coast. I use LiDAR surveys to rapidly survey large areas which are difficult to survey using conventional tools. After the data has been collected, it must be converted into a readable point cloud. Once I have imported the data into CAD, the engineers can investigate how coastal erosion is impacting the shoreline.

Summary

To sum up this area, think about how technology is used on almost everything you do. Consider how you have been able to make use of digital technologies, such as CAD or BIM.

2. Health, safety and welfare

Overview

The basic purpose of health and safety in the workplace is to ensure that all workers can perform their jobs in a safe and healthy environment. You should only carry out work if you have the relevant skills, knowledge, training and experience. You should be aware of the health and safety risks involved in your work and you should understand how those risks are managed.

EPA requirements

K5	Statutory health, safety and welfare policies, procedures, and regulations, including risk management, in relation to civil engineering project delivery	4(a)
S4	Apply statutory health, safety and welfare policies, procedures, and regulations in the civil engineering environment, using risk management processes, procedures, and documentation	4(a,b,c)
B1	Comply with health, safety and welfare requirements, industry standards, statutory regulations, policies and codes of practice	4(a,b,c)

The KSBs have been mapped across to the Attributes as featured in the ICE publication (ICE, 2021).

Interpretation

Use this section of the application to demonstrate how you use your knowledge about safety, apply it to your work and comply with policies or procedures. Help the assessors appreciate the type of well-defined rules you follow and demonstrate how you comply with health and safety associated with your work. Highlight some of the safety training you have completed so you can ensure you are equipped to work safely, understand the associated risks and are conforming to industry standards.

Sample responses

Contracting and construction: site supervisor – highway repair

I am responsible for making sure that the appropriate signage is used and placed in accordance with the Traffic Signs Manual. The risk assessments identify potential hazards. The necessary control measures are identified and I follow the steps outlined in the approved method statement.

Consultancy and design: assistant CAD designer – producing drawings

I work alongside the rail engineer to translate the design into buildable solutions. I use the guidelines for 'safe by design' provided by the client to highlight unusual and unmitigated hazards on my engineering drawings. I remove construction hazard triangles when producing as-builts.

Infrastructure owner/client: asset supervisor – water supply

I am responsible for planning work and ensuring the programmes for my projects are well defined. I hold weekly progress meetings. These meetings give me an opportunity to discuss any problems on the project. I inspect the works to check fixtures and fittings have been installed correctly and are compliant with the water regulations.

I am responsible for carrying out safety audits to check safe working practices are in place. I have the right knowledge of the requirements and suitable experience which means I can make sure both workers and the public are safe when street works are carried out.

Summary

Safety is a critical area for the civil engineering industry. Consider carefully how you define what you need to do to keep yourself safe and to work safely with others. As a civil engineering technician, you will be responsible for assisting in the preparation and production of plans in compliance with various codes of practice or industry standards, such as the Health and Safety at Work Act 1974, Construction (Design and Management) Regulations 2015 or the Design Manual for Roads and Bridges (DMRB).

Civil engineering technicians must comply with health and safety regulations, welfare and guidance on wellbeing. This includes not just physical safety but mental safety too. While you prepare your response to this area, consider how you contribute to a project or two but also consider some of the processes and procedures your company puts in place to protect you and look after your health, safety and wellbeing.

3. Project management

Overview

You may think you have little to do with project management. This is often seen as a role carried out by others. This area of the PoE is looking for you to demonstrate how you manage your time and how you identify what resources you need. How do you make sure you take responsibility for reducing error, maintain quality standards or simply get work done on time and to budget?

EPA requirements

K9	Project management, quality management and assurance systems and continuous improvement as applied to civil engineering	2(a)
S7	Plan, carry out and manage own work in line with quality assurance, recognising the wider implications to customer needs, and within cost and resource limitations	2(a)part 3(a)
S9	Apply document control processes and procedures using the approved processes, maintaining quality compliance when creating or amending engineering documentation	2(c)
B2	Work independently, operating in a systematic, proactive, and transparent way, using resources effectively to complete tasks, knowing their limitations and when to ask for support or escalate	2(b)

The KSBs have been mapped across to the Attributes as featured in the ICE publication (ICE, 2021).

Interpretation

At the very least, define how you take responsibility for planning and managing your own time. This may be looking about one or perhaps three weeks ahead to help you and others appreciate your workload. How do you know you are using the latest information? How is quality managed? How do you track progress of work done?

Providing a copy of your CPD records will go a long way to demonstrating your professional commitment. To be able to work as a professional takes time and enthusiasm. This is the behaviour you are looking to demonstrate that you have acquired as you conclude your apprenticeship.

Sample responses

Contracting and construction: setting out surveyor/site engineer
At the beginning of the week, I look ahead to see what work needs to be done. I check we have the right tools and equipment in the stores. At the end of my shift, I complete a site diary to record the work done, weather conditions, number of workers and any variations or problems encountered. Sometimes I include photographs. I upload all this information to the common data environment.

Consultancy and design: assistant CAD designer – producing drawings
Every week my team holds a resource meeting. I'm responsible for advising my line manager on my workload. If I have annual leave planned, I highlight this on our resource planner. Every day I record the work I have done in a timesheet. I am responsible for checking my drawings to ensure they are accurate and complete. I get my drawings checked by the engineer before they are issued.

Infrastructure owner/client: asset supervisor – rail maintenance
I was recently tasked with undertaking a survey of a bridge. To access the structure, I needed to work at height and use a mobile working platform. I raised this with my line manager and arranged for an operator to be available on the night shift to facilitate the movement of the platform. On the night I was responsible for checking their qualifications were current.

Summary
As a professional, you have a duty to manage your time and deliver work to a high standard. You should be taught this from the very beginning of your apprenticeship. Over time, you will become more skilled and, with sufficient experience, you should have the ability to work independently. At the conclusion of your apprenticeship, you should be a reliable member of the team.

You should be capable of following approved processes and, at the very least, comply with company quality controls related to the delivery of work or control of documentation. You should know what to do and where to go when you reach the limit of your knowledge or responsibilities. Think about what you are responsible for in your role: when can you take the initiative and when do you need to escalate things to others?

4. Personal and professional practice

Overview
By caring about the quality of your work, by making a commitment to stay up to date with industry trends and best practice, by being a reliable and trustworthy member of the team, you will be doing the right thing. Integrity and honesty are the very foundations of what makes someone a professional. To be successful in your EPA you need to be able to demonstrate how you behave like a professional.

EPA requirements

K11	Ethical principles as applied to civil engineering and the security of data and information	7(b)
K12	The values and standards by which they maintain their personal, professional, and technical knowledge and skills through initial professional development (IPD) and continuing professional development (CPD)	7(c)
S11	Apply ethical principles to civil engineering projects, including the secure use of data and information	7(b)
S12	Plan, undertake and review their own professional competence, regularly updating and reviewing their CPD to improve performance	7(c)
B4	Is motivated when collaborating in teams, offering sensible challenge, reflects on and provides constructive feedback and contributes to discussions	6(b,c)
B5	Maintains professional and ethical working relationships with internal, external, and connected stakeholders	7(a)part 7(b)
B6	Takes responsibility for their own professional development, seeking opportunities to enhance their knowledge, skills, and experience	7(c)

The KSBs have been mapped across to the Attributes as featured in the ICE publication (ICE, 2021).

Interpretation
All the requirements of this part of the portfolio could be viewed as being interconnected with the ICE Royal Charter. The ICE defines six rules of professional conduct. If you want to be a member you should take a look at them. There is more on these in chapter 4, Attribute 7.

If you apply for an EPA on its own or an EPA combined with EngTech MICE, you need to demonstrate your CPD (K12, S12 and B6) and ethical principles. You need to take responsibility for your professional development; plan, undertake and review your CPD. You need to care about your work and maintain professional standards and relationships with all stakeholders.

Sample responses

> **Contracting and construction: assistant site manager on refurbishment of a bridge over water**
>
> I am responsible for making sure we use the correct materials on site in accordance with the design engineer's specifications for the repair of the mortar joints in the masonry parapets. I make sure I record what work was done each day. I am also responsible for making sure we put the correct measures in place to stop materials falling into the river during the repair. By preventing spillages, I reduce pollution and this helps to protect the environment.

> **Consultancy and design: CAD technician – new nuclear facility**
>
> I am currently detailing the construction sequence for a crane on a nuclear facility. Before I could work on the project, I needed authorisation to work with classified information. I also undertook a range of training to understand the specific regulations and standards that govern nuclear facilities. I recorded my training in my PDR.

> **Infrastructure owner/client: asset supervisor – transport systems**
>
> I am responsible for identifying any training needs. Every year I have an annual appraisal with my line manager where we discuss my personal and professional goals. On a regular basis I undertake training on the latest cybersecurity threats. This helps me to be vigilant to things like phishing simulations or social engineering and minimise the risk of a data breach.

Summary

Most apprentices receive training at the beginning of their employment on subjects like data security, diversity and inclusion. As you progress through your apprenticeship you will use that knowledge and apply it to your working life. You will put your training into practice, keeping data secure and treating people professionally, with dignity and respect. Within your first year you will learn how to record your training. The tripartite meetings are a natural way for you to review your progress against the KSBs and plan how to improve your performance.

As your capabilities grow and your confidence develops, you will begin to contribute to team discussions related to your projects. You will have the ability to provide constructive, useful feedback. In addition, you will start to become more aware of the type of work you want to do and the skills you will need to develop. CPD is essential for maintaining competence and career growth. You will be actively involved in seeking opportunities to develop your career.

Level 4 portfolio of evidence

If you are on a level 4 apprenticeship you should look at the EPA requirements for a *senior* technician. The key difference lies in the complexity and depth of the work being done. At level 3 there is a focus on demonstrating basic technical skills and knowledge. In contrast, the level 4 PoE requires more detail to demonstrate the more advanced skills and responsibilities.

1. Design, technology and modelling in civil engineering

Overview

This is key to showing how you use computer-based software packages to control processes to carry out your responsibilities. This could include advanced use of BIM, such as coordination and integration of different models into a large common data environment. You should be able to explain and demonstrate how important digital modelling is to your role and how you can show the limitations based on your practical knowledge and experience.

EPA requirements

K6	Design principles and control processes used in the civil engineering consultancy, construction or manufacturing process, and the common constraints faced	1(a)
K7	Technical drawings, designs, and models, using analytical and computer-based software packages	1(b)
K8	Uses and limitations of computational and digital models, including building information modelling (BIM)	1(b)
S6	Produce and interpret civil engineering technical drawings, designs, and models, using analytical and computer-based software packages, recognising the limitations of the software used	1(c)

The KSBs have been mapped by the author across to the Attributes (ICE, 2024a). Mappings are indicative as there is no formal guide.

Interpretation

You will need to show how you apply your understanding of CAD or BIM in the production of designs or delivery of civil engineering. Below are some short examples which you could use to expand your ideas on how you may link your work to the EPA requirements. Remember to clearly define the problems you solve and link your claims to evidence to demonstrate what you are capable of doing.

Sample responses

Contracting and construction: setting out – drainage
I use the BIM model to get available information about nearby utilities. This gives an indication of the type of utilities and the probable location underground. I then check this information on site and do a CAT and Genny check to pinpoint more accurately the location of underground services.

Consultancy and design: CAD technician – producing drawings
I import data from the topographical survey and cross-reference it with the boundary survey and the highway engineer's road geometry. I then prepare drawings to show the construction information at different chainages. This helps the contractor work out the setting out on site.

Infrastructure owner / Client: Asset Inspector
I use vehicle mounted LiDAR equipment to rapidly survey large areas and gather huge amounts of data. On smaller projects I simply visit the site and carry out a visual inspection using more conventional tools such as a theodolite and measuring tape.

Summary

Think about how technology is used and the impact it has on almost everything you do. Consider how digital technologies are used to improve how we model, draw or deliver engineering. Also consider the limitations. Take the opportunity to explain how you use the software you are familiar with and how you apply it in your work.

2. Project management and safe systems of working

Overview

The main focus is on the different tools used to deliver projects in relation to managing programmes, quality and safety. We all are expected to work to deadlines; we do this by planning and organising ourselves and working with others. As we complete various items of work, we need to report on our progress and ensure we maintain good standards of quality.

In addition to being a reliable member of the team, you should be able to identify various different hazards and risks related to your work. You should be able to evaluate the risks and show how you contribute to eliminating hazards or reducing risks.

EPA requirements

K10	Statutory health, safety and welfare policies, procedures and regulations, including the Construction (Design and Management) regulation	4(a)
K11	Risk assessment and mitigation processes, and their importance in the civil engineering environment	4(b)
K13	Project management techniques, including quality and information management and assurance systems and continuous improvement processes	2(a,c)
S8	Comply with, and encourage others to demonstrate, statutory health, safety and welfare policies, procedures and regulation	4(c)
S9	Complete risk assessments to identify, evaluate and mitigate risks	4(b,c)
S11	Employ project management techniques, measuring and recording progress against civil engineering project plans	2(a) 3(a)
S12	Assess and report on quality using appropriate management and assurance systems and continuous improvement processes	2(a)
B1	Works to health, safety and welfare requirements, industry standards, statutory regulation and legislation, policies, and codes of practice, and ensuring others do likewise	4(a,b,c)

The KSBs have been mapped by the author across to the Attributes (ICE, 2024a). Mappings are indicative as there is no formal guide.

Interpretation

Use this section to demonstrate how you use your knowledge of Construction (Design and Management) Regulations 2015 (CDM 2015) and the procedures and policies in your organisation to contribute to the management of health and safety. Show how you apply safe working practices to your work. Help the assessors appreciate the type of well-defined rules, such as risk assessment, method statements, hazard identification or safety checks, you follow to demonstrate how you comply with associated health and safety regulations.

Highlight some of the safety training you have completed so you can ensure you are equipped to work safely, understand the associated risks and are conforming to industry standards.

Sample responses

Contracting and construction: site supervisor – highway repair

I have the knowledge and skills to carry out cable avoidance work safely. My certificate is valid for three years. I am responsible for issuing 'permits to dig'. This is part of my company's safe system of work to comply with health and safety regulations. This helps identify the hazards and reduce the risk.

Consultancy and design: assistant CAD designer – desk based

I am responsible for continually assessing my workstation and reporting any issues. On an annual basis, an assessment of my display screen equipment is undertaken and recorded. This helps to reduce any repetitive strain injuries. I take short screen breaks to reduce the risk of eye strain.

Infrastructure owner/client: asset supervisor – water supply

I am responsible for managing the programme for the works. I ensure the water supply is protected from contamination. I hold a valid CSCS card and I have a 'Blue Card' from EUSR in water hygiene. I have a NRSWA street works supervisor qualification and follow the 'Red Book' which is an ACOP for street works to help me put in place the right signage, lighting and guarding.

Summary

When putting examples together, consider how you successfully deliver your work and how you need to take into consideration how projects are managed and progress is monitored. Show how you complete your work with due consideration for quality and safety. Show how you have the proper training and information about relevant health and safety practices to protect yourself and look after your health, safety and wellbeing. You may be able to get your application to stand out by reflecting and evaluating how changes to health and safety legislation have benefited the industry.

3. Roles, responsibilities and engagement with others

Overview
As a senior apprentice, you will have developed the ability to coordinate and manage your work on projects. You will be in a position where you are not just supporting the work of engineers but, when appropriate, you will be capable of working in parallel with them.

EPA requirements

K16	Roles and responsibilities within the organisation, team dynamics and their own boundaries of authority	6(b)
K17	Relationships between key organisations in the civil engineering sector (for example organisations, customers, partners and suppliers)	6(b)
K18	Equality, diversity and inclusion, its importance and impact on civil engineering solutions	6(d)
S14	Monitor and manage individual performance, and supervise others, recognising the need to comply with appropriate codes of practice and equality, diversity & inclusion (EDI) requirements	6(d)
B3	Works individually and as part of a team, being aware of their actions and the impact they may have on others, and demonstrating awareness of diversity and inclusion issues so as to meet the requirement of fairness at work	2(b) 6(b,d)
B5	Maintains professional and ethical working relationships with internal, external, and other stakeholders	6(a,b) 7(b)

The KSBs have been mapped by the author across to the Attributes (ICE, 2024a). Mappings are indicative as there is no formal guide.

Interpretation
We all work in teams. You should look to describe one or two of the teams you work in. These teams could be made up of just a couple of people or they could be much bigger. The people could all be in the same organisation as you or they could include people from outside your organisation – for example, clients, consultants and contractors.

Use this section to identify the different roles and responsibilities of the different stakeholders you work alongside. Who do you report to? Describe their roles and the different levels of responsibilities they carry. Describe how you communicate with them and others and the importance of good communication. Explain how you monitor and manage your work. What are you responsible for planning and managing? How do you track progress of work done?

Describe how your organisation supports fairness at work. How do they encourage diversity and inclusion? What are your responsibilities?

Sample responses

> ### Contracting and construction: setting out surveyor/site engineer
>
> During the summer I was responsible for supervising a summer student from university. I provided an overview of the project, site safety rules and helped them settle into their role. I gave the student specific tasks to complete that matched their skill levels and learning objectives. At the end of the summer, I was responsible for reporting on their progress to the construction manager.

> ### Consultancy and design: trainee CAD technician – producing technical drawings
>
> I'm responsible for advising my line manager on my workload. I need to keep my team informed of progress on my projects. Every year my company holds annual appraisals of all staff. These help to assess achievements, identify key areas of improvement and set goals for career development. This year, on successful completion of my apprenticeship, I hope to get promoted to CAD technician.

> ### Infrastructure owner/client: asset supervisor – rail maintenance
>
> I am part of a team of inspectors. We come from a range of backgrounds. My expertise is in masonry. I work night shifts which helps me find time to meet my children's needs. At the end of a shift, I am responsible for reporting my findings to my shift supervisor. This helps maintain a safe and efficient railway.

Summary

You will have your own set of values. As a professional, you have a duty to always act with diligence and care. Ethics are about being honest. By the end of your apprenticeship, you should have become a reliable and trustworthy member of the team. Being in a team is more than just getting work done, it's also about treating people with dignity and respect.

When you are gathering data together, think about what evidence you could use to demonstrate that you work in a diverse team. Take the opportunity to show what you know about equality, diversity and inclusion. Think about how you maintain professional working relationships with your colleagues, clients and other stakeholders.

4. Personal and professional practice

Overview

Personal and professional practice is focused on the way you develop and maintain your competence and knowledge. A key focus will therefore be on how you carry out and record your own CPD necessary to maintain or enhance your abilities. You should also highlight how you support others. Good ethics involve the principles of honesty, integrity, fairness and accountability. By becoming a reliable and trustworthy member of the team, you will become dependable. To be successful in your EPA, you need to be able to demonstrate how you behave like a professional by caring about the quality of your work and taking responsibility for continuously seeking to improve your knowledge and skills.

EPA requirements

K19	Ethical principles as applied to civil engineering, including the need for the confidentiality and security of data and information	7(b)
K20	Methods to maintain professional competence and technical knowledge, including initial professional development (IPD) and continuing professional development (CPD)	7(c)
S16	Apply ethical principles to civil engineering projects, including the secure use of data and information	7(a,b)
S17	Plan, undertake and review their own professional competence, updating and reviewing their CPD to improve performance	7(a,c)
B6	Takes responsibility for their own professional development, seeking opportunities to enhance their knowledge, skills, and experience, and support others when requested	7(c)

The KSBs have been mapped by the author across to the Attributes (ICE, 2024a). Mappings are indicative as there is no formal guide.

Interpretation

All the requirements of this part of the portfolio could be viewed as being interconnected with the ICE Royal Charter. The ICE defines six rules of professional conduct. If you want to be a member you should take a look at them. There is more on these in chapter 4, Attribute 7.

Apart from cyber attacks, data security also involves protecting against physical theft or loss of devices, ensuring data storage and backup, and controlling access to sensitive information. What measures do you take to keep data safe? A key way to ensure data integrity and confidentiality is through regular training to maintain knowledge and awareness of the risks. How do you receive this training?

Professional development is an ongoing process that adapts to the evolving needs of the workplace, whereas the academic education you get from school or college provides the foundational knowledge and skills on which professional development builds. If you apply for an EPA on its own or an EPA combined with EngTech MICE you will need to demonstrate how you do CPD (K20, S17 and B6). You need to take responsibility for your professional development; plan, undertake and review your CPD. You need to care about your work and maintain professional standards and relationships with all stakeholders. You should also encourage and support others to hold the same values.

Sample responses

Contracting and construction: assistant site manager on refurbishment of a bridge over water
I take great care to ensure all the data and information I use on the project are accurate and up to date. I review and verify the information provided to avoid errors. I have recently refreshed my training of the Bribery Act through an online training course. This has reminded me how to identify and avoid risky situations. A student on their industrial year started on site this summer. I was asked to supervise them, be supportive and provide guidance on day-to-day tasks.

Consultancy and design: CAD technician
Every year my line manager conducts my annual appraisal. This includes a skills scan to identify my strengths and areas for improvement. This helps me plan my CPD. Last summer, I had the opportunity to assist a school student who was on work experience. I helped them to settle in and I shared my knowledge about civil engineering and the different careers that are available.

Infrastructure owner/client: asset supervisor – transport systems
I often update my training on the latest cybersecurity threats. This keeps me informed about new attack methods and how to handle them. In the future I hope to become a contractor's responsible engineer (CRE); to do this I need to first enrol on a civil engineering degree programme. I have spoken with my line manager, and I have put this as a goal in my development action plan.

Summary

The primary focus for this area is on continuing professional development. At the start of your career, you are likely to have limited understanding of professional development. As your knowledge and experience develop, you will learn how professional development will help you to adapt and stay up to date with industry trends. Professional development is less rigid and formal when compared to formal qualifications. The outcomes are more often focused on immediate application and career growth.

When you are gathering data together, think about what evidence you could use to demonstrate what you now understand about professional development compared to what you knew about training and development at the beginning of your apprenticeship. CPD is essential for maintaining competence and career growth. A significant part of this area of the application is about how you plan and manage your ongoing training and development and how you encourage others.

Supporting evidence

The portfolio of evidence for either level 3 or level 4 should typically contain 10 to 12 individual pieces of discrete evidence mapped against the KSBs (see Figure 7.3). The supporting

Figure 7.3 Examples of appendices (author's own)

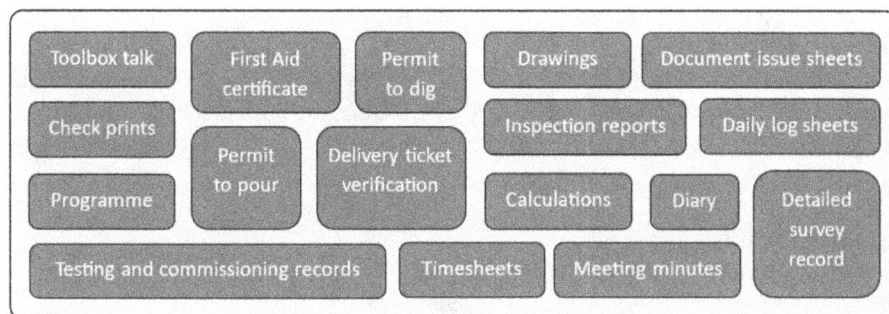

evidence is simply proof that you have done what you claim you have done. The 'perfect' evidence will have a project name, your name and the date you were involved written on it. This is ideal for some evidence, such as drawings, emails, diary sheets or any document that forms part of a quality process.

Think about what you need to do in your ordinary job. How do you prove you did a test on site? How do you demonstrate you have carried out a check or you have followed a process? What do you sign, what do you put your name on? What do other people put your name on? All these things will be evidence you are doing work and taking responsibility for the work you (and sometimes others) do.

Don't just think about the product of your work. There will be evidence of others putting your name on things, such as resource charts. If it is not clear how the evidence relates to you and your role, consider including a short sentence to clearly state how it is relevant to you.

Carefully consider each piece of evidence you choose: does it add value? Does it help prove to the assessors you are now a professional working at the same level as EngTech MICE? Does it show them what you can do or need to do to make sure your work is done to a high standard? Any evidence you submit is open to scrutiny so be prepared to discuss it and answer questions about it in more detail.

Sponsor's statement of support

When applying for professional recognition with your end point assessment (EPA) you will need to choose *two* people to sponsor your application. Sponsors must meet certain requirements, so it's important that you read up on who is eligible to be a sponsor and what they are required to do.

You must nominate one of the sponsors as the lead sponsor. The lead sponsor must be recognised as an MICE/FICE. Both sponsors must upload their statements before the ICE deadline; check online for more details (https://www.ice.org.uk/join-ice/key-membership-dates). The statements should be uploaded directly to the ICE portal (https://reviews.ice.org.uk/upload/epa/UploadQuestionnaire). The documents should be in pdf or MS Word format.

Lead sponsor

The lead sponsor plays a crucial role in supporting candidates throughout the EPA process. They must be a member of the ICE. The Institution expects the lead to be familiar with the current requirements. The lead sponsor must complete all sections of the statement of support and provide a statement to describe you, your skills and your abilities. They should describe from their own personal knowledge why they consider that you possess the requisite abilities and characteristics and that you are a 'fit and proper person' suitable for admission to membership of the Institution (ICE, 2024e).

Second sponsor

The second sponsor has little more to do than simply confirm, by provision of their signature, that you possess the skills and characteristics comparable to an EngTech MICE and that you are a 'fit and proper person' (ICE, 2024e). If they wish to, and I would encourage it, they can also provide a short statement of up to 500 words, but this is not mandatory.

Continuing professional development

As part of the professional discussion on your EPA day, you will need to demonstrate how you plan, undertake and review your professional competence. You will need to show how you regularly update and review your CPD to maintain and improve your performance.

The easiest way to do this is to submit a full copy of an ICE continuing professional development (CPD). This combines a current development action plan (DAP) and an up-to-date professional development record (PDR). CPD is defined by the Institution as the 'systematic maintenance, improvement and broadening of knowledge and skills, and the development of personal qualities, necessary for the execution of professional and technical duties throughout your working life' (ICE, 2024d). The Institution's rules state that CPD is mandatory for all members.

The Institution normally requires you to submit three years of records, although one year is acceptable. If you do not have this, you should provide an explanation in your submission and, if asked by your assessors, be able to show how you have maintained your competence. This leaves it all to chance on the day. It's better to take a little time before you apply and build up a suitable quantity of evidence and submit sufficient records to demonstrate you have undertaken at least 30 hours' effective learning per year for the three years, if experience allows, prior to your application.

Development action plan (DAP)

In the DAP you should identify your future goals. When setting your goals, they should be specific, measurable, achievable, realistic and timely (SMART). The ICE template will help. While in principle this is a good idea, keep in mind that it's not always possible to define specific objectives and so it is ok to put in some nonspecific, open-ended goals too.

Professional development record (PDR)

The PDR is a detailed record of learning activities and knowledge gained. The requirement for a PDR can initially cause some alarm. But it needn't. As an apprentice you are likely to be preparing for your EPA after a long period of training. While under training you will have had to produce a regular record of your training achievements in your tripartite meetings, recording both off-the-job and on-the-job training and ultimately had this recognised in your Gateway training review.

The ICE CPD guidance covers various types of training beyond traditional courses and certificates. It includes structured learning (such as attending workshops), self-directed learning (reading or watching recordings), informal learning (participating in webinars), on-the-job learning (gaining experience through hands-on work projects) and peer learning (engaging with colleagues). It is good to have a diverse range of learning opportunities recorded in your PDR.

Submitting your application

Before you submit your application, you should get it checked by someone who understands the requirements, such as your company mentor. Then submit the whole application to your training provider. They will conduct a final check and upload it. The application is submitted by way of the ACE 360 portal to the ICE. You do not have to pay for your EPA. The cost will be paid by your training provider.

If you have prepared well and taken time to plan this part of the process, identify the tasks that need to be done and organise your time effectively, you will probably be able to put your application together within three or four weeks of passing through Gateway. It is good practice to keep in communication with your college throughout this period so that they are aware of your plans and ambitions. It is good practice to consider giving your college 5–10 working days to independently check your application before they upload it to the ICE deadline.

Receipt of application

The ICE will assess your application, ensure that all necessary documents are in order, verify the completeness and assess its eligibility. If it is deemed valid, they will acknowledge receipt within ten working days. Shortly after this you will be told the details of the EPA and you will start final stage which is related to the technical project brief.

If you do not get a response within 10 days, first double check your junk folder, then politely write to the end point assessment team at epa@ice.org.uk to remind them of the promised response time and ask for an update.

Responding to the technical brief

You will be told the details about the EPA at least six weeks before the EPA date. The notification will give you the names of your assessors and the timings for the day. If your EPA is going to be online, you will receive a formal meeting request from an ICE staff member. The meeting request will contain a Microsoft Teams link for your interview. If you do not receive a meeting request, you should contact the end point assessment team directly.

You will be provided with a *scenario* and *challenge* from the ICE related to the area of technical specialism and the scenario code your employer selected as part of your application. The ICE provides some guidance explaining the requirements, the format for the report and presentation and a reminder of the marking scheme.

You will have *six weeks* to complete the project report and presentation. The level 3 guidance indicates it will take you typically up to *30 hours* of work over the six working weeks to complete your response (ICE, 2024g). The level 4 guidance remains silent on the hours required so you are left to decide for yourself on what would be suitable. Roughly speaking, 30 hours of work is considered (so far as an apprenticeship is concerned) to be equivalent to either a full working week or one day a week for five weeks. How you choose to split your time will be down to you and your employer.

Did you know?

The technical project report, presentation and question-and-answer session on the day only cover some of the KSBs.

Level 3 KSBs:

Knowledge: K1 K2 K3 K4 K5 K6 K7 K8 K9 K10 K11 K12

Skills: S1 S2 S3 S4 S5 S6 S7 S8 S9 S10 S11 S12

Behaviours: B1 B2 B3 B4 B5 B6

Level 4 KSBs:

Knowledge: K1 K2 K3 K4 K5 K6 K7 K8 K9 K10 K11 K12
K13 K14 K15 K16 K17 K18 K19 K20

Skills: S1 S2 S3 S4 S5 S6 S7 S8 S9 S10 S11 S12
S13 S14 S15 S16 S17

Behaviours: B1 B2 B3 B4 B5 B6

(IfATE, 2021b, 2022, 2024a, 2024b)

Interpretation

As the title suggests, your report should be a technical report. It should bring out some of the engineering principles you have been taught at college or are now familiar using in your daily work. You should use the technical report to demonstrate some of the appropriate engineering methods and techniques you use to contribute to the design, construction or maintenance of infrastructure and buildings.

What do you know about civil engineering and the construction process? What materials do you have experience of using? Do you know the correct engineering terminology related to your field of expertise? Other key skills to consider are the impact that diversity and inclusion can have on the delivery of the project and how you take these matters into consideration.

In addition, you should identify some of the software you use to deliver engineering solutions. This could be related to either the design or construction process. How do you ensure the work you do is done correctly? Finally, you will demonstrate how you use your knowledge and skills to communicate by writing a technical report, preparing a presentation and responding to the Q&A.

Did you know?

The civil engineering technician apprenticeship requires you to complete a technical project based on a *real, work-based* civil engineering *challenge*.

You will respond by

- producing a report
- producing a presentation
- preparing for a question-and-answer interview session on the project.

Technical project report

The technical project reflects a typical challenge that you could be involved in delivering as part of your ordinary working day. You will be given a *scenario* and a *challenge*. You will need to use a real-life project to base your report on. You should work with your mentors to identify a suitable project. You will need to work closely with them to make sure your report meets the brief and demonstrates the appropriate KSBs.

In this section we will provide a short guide to give you and your sponsors some hints and tips on how to approach the technical project report. The aim is to help you get started, give you some direction and enable you to confidently deliver a suitable response. In addition to the report, don't forget you also need to create a 10-minute presentation and be prepared for a 20-minute Q&A session.

The technical project report for level 3 must be 2500 words ± 10% and for level 4 it must be 3500 words ± 10%. It should have a professional layout with a table of contents. The appendices, references, diagrams and so on will not be included in the word count. When preparing the report, consider your reader. They are likely to be reading this document on a laptop or tablet. But it is also possible they may print it out. Hyperlinks could be used to link between different sections of the report, but not to websites or other items outside of the report itself.

IfATE (2021a) provides a guide for what should be included in the technical project report. In addition, both sets of guidance (ICE, 2024g; ICE, 2004a) have a summary of what must be included. Below is a suggested structure for your report, which could suit either level 3 or level 4 EPA, which you may wish to adopt or adapt.

	Level 3	Level 4
• Cover sheet		
• Introduction	100 to 150 words	100 to 150 words
• Objectives	200 to 250 words	200 to 300 words
• Project plan	200 to 250 words	250 to 350 words
• Research and results	1000 to 1,200 words	1600 to 1800 words
• References	200 to 250 words	250 to 350 words
• Project outcomes	250 to 300 words	350 to 450 words
• Conclusions	300 to 350 words	400 to 500 words
• Appendices		
	2250 to 2750 words	3150 to 3850 words

Cover sheet

Include a title, space for a photograph (for identification purposes) and space for your name and your personal identification number. You should include space for a signature from your lead sponsor. They must verify that the report is *all your own work*. You may consider using an image to illustrate your project brief. A good cover sheet will look professional and provide a good first impression.

Introduction

For your introduction you should give the background to the project and explain the main aims. You could expand on this and show how the scenario and challenge you have been set relate to the 'real, work-based civil engineering challenge'.

Objectives

Take the opportunity to fully define the *scope of the project* scheme and define the element you have selected to focus on – for example, a railway crossing (bridge/tunnel), development of a new building, construction of a new road, maintenance of a bridge over a river, or inspection strategy for an existing highway. There are a diverse range of different scenarios and challenges available. Be clear in the specific goals and define the project aims. At the same time, define the limitations of the project and set boundaries to enable you to focus the report.

Project plan

Outline the *key performance indicators* and timeframe for the project and define timelines and milestones. Consider the stage of the construction process you have focused on. For example, is it concept, tender, detailed design, construction, operation, maintenance or decommissioning? Think about the *plan, methodology* and *resources* you will be using in terms of time, materials, equipment or software and so on to deliver the solution.

It is a good idea to start with a 'kick-off' meeting with your mentors, just like you would on any ordinary project. This could be an excellent way for you to discuss the project, define the outcomes, the timeline (schedule, milestones and deadlines) and set out any potential issues – for example, whether you or your mentors have any planned holidays!

A well-conducted kick-off meeting will help ensure a smooth start and that you and your mentors understand the goals and how to achieve them.

Research and results

This is one of the main sections in the report. This is where you should focus on collecting relevant data, undertake appropriate analysis and record your *findings*. You may be looking at two options to cross a railway. Or you could be looking at the maintenance of a bridge or construction of a highway.

The assessors want to know what standards you work to – for example, DMRB. Do you have an engineering mindset? This will require you to undertake some *analysis and evaluation* of the results. The analysis should be in keeping with the responsibilities of the apprenticeship linked to EngTech MICE, *not* the level of an engineer such as IEng/CEng MICE.

You could be asked to calculate the volume of an embankment or excavation. Depending on your role, you may add value to this by converting the volume into a tonnage, illustrating the different materials or layers in the volume or estimating the number of trucks you would need to remove (or deliver) the material to site. If you are proposing a repair to a bridge/road/structure, you may look at how you should prepare the existing material, what type of material should be used and how to apply it.

You should look to gather *data* together to demonstrate to the assessors that you are familiar with the engineering associated with your role and employment. This work could cover several pages and may be a significant proportion of the report. You may also include some calculations or technical data in your appendices to support the work you do in this section.

References

You should make reference to relevant *scientific and engineering principles*. Link to *relevant methods* of construction or design and *techniques* used. Highlight any limitations or considerations that could impact your project. It can help to identify *relevant industry standards, policies, regulations, legislation or guidance* which may be specific to your specialism in railways or highways.

You should explain any analysis you have done. If you are comparing two different options, then at this stage you should state which is your preferred option and why. This could include factors such as cost, efficiency, environmental impact or other factors. It will be beneficial to include any *environmental and sustainability concerns* that could have an effect, such as minimising waste, disposal of contaminated materials, managing dust, noise and vibration pollution during construction or designing accessible infrastructure for road users with different mobility needs.

Project outcomes

The project outcomes should demonstrate the practical application and impact of your research and findings. You should provide specific results and achievements that meet the project's objectives. Did you hit the planned targets? This section should be a factual and succinct section of the report.

Conclusion

This is the final part of the report. It is where you should wrap things up. You could include a summary of the main points and your findings. You should provide recommendations for the preferred or proposed engineering solution.

In your conclusion you should also provide an evaluation of your performance to determine the challenges you faced and how you overcame them. This could include an evaluation of the original project plan as agreed in the kick-off meeting. Did you hit your goals? Were there any unforeseen challenges? How did you resolve issues?

Appendices

Include any reference materials. There are no mandated requirements of what could be included. All appendices of supporting evidence must be attributable to you in full. They must be clearly referenced and labelled in your technical report. Example appendices of support-ing evidence may include plans, diagrams, calculations and designs. This is not a definitive list and other sources of evidence are permissible. It is suggested you keep your appendices succinct and aim for approximately 10 to 12 individual pieces of either A4- or A3-sized docu-ments of evidence. You may find it beneficial to map your appendices to the relevant KSBs for this assessment method. It may help you focus on what you are trying to demonstrate.

Overall, you should make sure your report is clear and concise. Although it is not necessary, you may choose to highlight the KSBs you are targeting, either in the margin of the report or at the footer of each page. This will keep your mind focused and make it easier for the assessors to read. If you include images, diagrams, graphs and charts you will find it easier to illustrate any key points. And make sure you proofread your report to check for errors. This will ensure your report is professional.

Did you know?

Once complete, you need to prepare a witness statement confirming the report is your own work, signed by both you and your employer. For example, they could include the statement:

"I confirm that the following report is all the candidate's own work."

Presentation

As part of your response to the technical project brief you must prepare a 10-minute pres-entation summarising your technical report. The slide deck should be clearly laid out. When preparing your presentation, you need to remember you are telling the assessors about your technical brief and your response to it. Like your report, your presentation should have a beginning, middle and end (see Figure 7.4).

The guidance indicates that your presentation must include a summary of your technical report (ICE, 2024g). Therefore, you may find it beneficial to use the structure of your report to form the basis of your presentation. Most people put together a PowerPoint-style presentation. The recommended time for each slide in a presentation varies depending on the content and detail in the slide. A common guideline is to use one slide per minute plus one for the introduction. Therefore, you are likely to need between eight and ten slides.

Keep in mind that you will be assessed on your ability to communicate clearly, incorporating relevant and appropriate information. You should lay out your presentation using a balance of words and graphics to explain your ideas. When embedding images, make sure they are clean, clear and easy to read.

Don't forget to make pdfs of the slides and bind them to your report when you submit your response to the technical project brief.

Submitting your response

Your technical project report and presentation must be uploaded within six weeks of receiving your technical project brief. The submission date will be detailed in your acknowledgement letter. Ideally, this is 3 weeks before your EPA date. Your *report* and *presentation* should be uploaded to the ICE EPA portal (https://reviews.ice.org.uk/upload/epa/ProjectBrief) and submitted within a single pdf file of no more than 15MB.

The end point assessment day

On the day, there are two assessment methods. The first will be based around your response to the technical project brief. The second is based on a professional discussion of the portfolio of evidence and, if applying for EngTech MICE, the CPD records you have submitted.

On the day, your assessors will introduce themselves. The lead assessor will set the scene for your EPA. They will explain how the EPA is carried out, including reminding you that they are there to engage with you and find out what you know. They are not trying to catch you out or ask questions that you cannot answer.

Whether you conduct your EPA online or in person, it is recommended you dress professionally. Choose clothing that you would normally expect someone to wear for a formal business meeting.

Figure 7.4 Presentation structure

Beginning:	Middle:	End:
Introduction to project	Research and findings	Conclusions
Scenario and challenge	References	
Project plan	Project outcomes	

Online

If you have indicated you would prefer to do your EPA online, make sure, prior to the day, that you have thought about when and where you will sit. Wherever you choose, make sure you have a good internet connection and you will not be disturbed by anyone.

Most people will routinely use Microsoft Teams as part of their normal day and therefore will be highly familiar with how it works. But this is not the same for everyone. If you are not familiar with it and you want to do the EPA online, then the ICE recommends you to have the following:

- laptop or desktop computer
- headset or microphone and speaker
- webcam
- internet connection.

You will need to get hold of the correct software and you should make sure it works on the machine you are going to use. It would be good practice to test your equipment (camera, microphone and speakers) in advance of the day. This reduces the risk of being distracted by last-minute audio problems when your thoughts would be better focused on the EPA. Bear in mind that Microsoft Windows issues regular updates to the system, which can affect the quality and stability of your MS Teams link, so check for Windows updates the day before your EPA.

You should join the meeting, by way of the link in the formal meeting request you were sent, at least ten minutes before the start of your EPA. You will be admitted into a lobby and you should wait for a member of the ICE staff to invite you into the main session.

Did you know?

If you are being interviewed online you must demonstrate that you are in a room on your own, remove any backgrounds, and you will not be allowed to blur your camera.

On the day there will be some preliminary checks. You will be required to confirm that you are alone, that your phone is switched off and show your proof of identity. So, make sure you have a passport, driving licence or similar document within easy reach. Once the checks are complete you will be asked to share your presentation and when ready you should make a start.

In person

On arrival at the venue, you should make your way to the reception desk, where your identity will be verified. After this you will be directed to the waiting area. At the appropriate time one of the assessors will come out to the waiting area to get you. During the short walk to the interview table your assessor will talk informally. This is not part of the EPA; be assured it has not started yet.

You will be given the opportunity to get yourself comfortable and set up your presentation. On the table there will be some paper, a pencil and some water. When it's time to start you will be asked to begin your presentation.

Assessment method 1

This assessment should take 30 minutes to complete. You will start by giving a 10-minute presentation to your assessors summarising your technical report. This will be followed by a 20-minute question-and-answer session. The assessors will have already read your technical report and seen the pdf of your presentation. They will not know what you are going to say or how you are going to say it. For the level 3 EPA the presentation and the question-and-answer session will be recorded using MS Teams if your EPA is online or audio-capturing equipment if your EPA is in person.

Presentation

Your presentation should be clear, concise and complete within 10 minutes. The presentation should be closely linked to your technical project report. Speak clearly and steadily; remember to look your assessors in the eyes. Also do not forget to smile. Try to look like you are pleased to be there. The more you practice, the more comfortable you will become. During the presentation the assessors will not say anything unless you go over your time limit.

The presentation is followed by a 20-minute question-and-answer session. This is intended to explore your response to the technical project brief. The assessors are not trying to catch you out or trip you up. They will have read your technical project report and listened to your presentation. They will want to find out more about how you arrived at your solution.

Online presentation At this stage, you will have prepared a good script and prepared the visual slides. The only thing left to do is to think about how the live performance will go. Keep in mind your assessors' perspectives and how they will be engaging with you in the question-and-answer session afterwards.

When presenting online to your assessors, make sure you are facing the camera, look down the lens and try to relax. You may find it more comfortable to look a centimetre or two above your camera. Looking at the camera mimics direct eye contact with your assessors. By using this approach, it is clear you are addressing the two assessors. This makes you appear more engaged and professional, which is better than if you were looking at the screen.

In-person presentation The same rules of engagement should be considered for the presentation as outlined in the online section. You should consider the scale of your presentation. Keep in mind that you are speaking to two people across a table and therefore a practical consideration would be to use either A3- or A4-sized documents. If you choose to use a laptop, make sure it is fully powered – the ICE does not guarantee access to a power supply.

Whether you do a presentation online or in person, all you are doing is talking about your response to the technical project brief. If you take time to focus on the purpose of the ten-minute presentation, then you may find it easier to approach. Done well, it will benefit you and the assessors in starting the question-and-answer session.

Question-and-answer session

There is a slight difference between the level 3 and level 4 presentation and question-and-answer session. Below is an interpretation of the KSBs which the assessors look for.

Level 3 presentation and question-and-answer session The assessors are looking to find out whether you can apply appropriate technical knowledge to solve the technical project brief, which should be related to a reasonably practical real-world project, such as designing or building something new or maintaining an existing piece of infrastructure. They want you to show how you can undertake some analysis to support the delivery of the civil engineering project brief.

You should also demonstrate how you use your knowledge of equality and diversity, health and safety, and environmental and sustainability policies to contribute to the solution – for example, by improving accessibility, eliminating or reducing hazards, or reducing carbon or material wastage. The question-and-answer session is another way you can show how you communicate.

Be mindful of the KSBs that you are looking to demonstrate. You should show how you can make decisions by applying known solutions based on clear guidelines to solve straightforward problems. Your response should be clear and coherent.

If you want to stand out, then you should evaluate your work and validate some of the assumptions you have made about your engineering analysis in relation to your project solution. Don't forget that the presentation and question-and-answer session for level 3 are recorded.

Level 4 presentation and question-and-answer session In addition to the practical skills found at level 3, you need to show a broader understanding of civil engineering beyond basic principles. You need to use some more advanced mathematical analytic techniques to demonstrate the greater knowledge you have gained and use it to contribute to solving the engineering problem.

You should be able to explain choices of materials or components related to civil engineering associated with the technical project brief. This could link to reducing carbon use or material wastage. You should be prepared to discuss some of the potential costs and the impact of safety, quality, the environment and security on your project.

This will enable you to demonstrate how the solution to the problem is not straightforward and that there are multiple variables that need to be considered. This should demonstrate how your decision making has developed and that you are able to take some initiative now you are ready for senior technician role.

Show how you planned the project and delivered it. Explain how you sought support to help you deliver the response. Finally, use the report, your presentation and the question-and-answer session alongside any appropriate technology to communicate clearly to your assessors. The level 4 assessment expects a higher level of critical thinking.

To aim for a stretch target, you should evaluate the information you have used, the techniques you have put into practice and justify how they helped you to solve the problem. You should look to evaluate your own performance when managing the project: did you deliver to the original programme? Could you improve your performance? Similarly, is there any way you could improve your response to further reduce the impact on the environment?

Assessment method 2

The second assessment method is based on a 40-minute professional discussion underpinned by your portfolio of evidence and, if applying for EngTech MICE, the CPD records you have submitted. You will be expected to be able to discuss in a professional manner the work you do and how you do it. Below is an interpretation of the KSBs which the assessors look for in a level 3 or level 4 EPA.

Level 3 professional discussion

The assessors will explore what software packages you use; what health, safety and welfare issues affect your field of civil engineering; how you manage your workload; how you deliver good quality work and how you consider diversity and inclusion and maintain good working relationships. You are encouraged to refer to your portfolio when answering questions.

Be mindful of the KSBs that you are looking to demonstrate.

They will also want to know how you plan your training needs, review your CPD and take responsibility for seeking opportunities to enhance your skills. This requirement applies whether you seek EngTech MICE or not. This is identified in KSBs K12, S12 and B6. By providing a copy of your CPD records, you are giving the assessors an indication of how you plan and review your own professional development.

Level 4 professional discussion

The assessors will explore what software packages you use and how you analyse engineering problems. They will want you to explain some of the limitations you have encountered, such as compatibility issues when sharing files between different software packages. You should be able to discuss some of the important health, safety and welfare regulations that affect your field of civil engineering. You should discuss how you comply with industry standards and ensure that others do the same. You should take the opportunity to talk about the teams you work in and explain how you manage your workload and monitor progress. Explain how you follow quality assurance processes to minimise errors in your work and how quality is controlled to ensure what you produce is to the right standard. Finally, demonstrate your awareness of the impact of your actions on others and explain how you build and maintain good working relationships. You are encouraged to refer to your portfolio when answering questions.

Be mindful of the KSBs that you are looking to demonstrate.

They will also want to know how you plan your training needs, review your CPD and take responsibility for seeking opportunities to enhance your skills and support others. This requirement applies whether you seek EngTech MICE or not. This is identified in KSBs K20, S17 and B6. By providing a copy of your CPD records, you are giving the assessors an indication of how you plan and review your own professional development. The assessors will also want you to demonstrate how you support others too.

> "The day was very nerve-racking, but I was very confident in the preparation and knew I was as ready as I would ever be."
>
> **Ayomikun Ajayi EngTech MICE**

Observers

It is worth noting that there may be an observer at your EPA. This can happen in either online or in-person EPAs. If there is a fourth person attending, you will be notified on the day. The observer is present solely in an observational capacity and does not actively participate in the EPA process. They are usually an auditor or being trained as an assessor. If online, you will notice an additional face on the screen. At an in-person EPA, the observer will sit just within your field of vision at one end of the table. They will not participate in the assessment process.

Summary

The EPA is your opportunity to showcase how professional you are, now you have completed your training. Overall, you need to demonstrate you have an appropriate balance of knowledge, skills and behaviours. The EPA gives the assessors the opportunity to verify that you have met all the required standards and competencies outlined in the EPA assessment criteria.

Table 7.2 below summarises the similarities, differences and details associated with the two different levels of EPA assessment day at the Institution of Civil Engineers.

Table 7.2 Summary of details of the two different assessment days

	EPA level 3 v1.1 (CET)	EPA level 4 v1.1 (CEST)
Part 1: 30 min	10 min on the TPB	10 min on the TPB
	20 min questions on response to TPB (recorded)	20 min questions on response to TPB (not recorded)
Part 2: 40 min	40 min questions on remaining KSBs	40 min questions on remaining KSBs (includes greater emphasis on BIM and evaluation of digital modelling techniques)
Results:	Distinction / Pass / Fail	

The assessors will ask open questions and try to gain an understanding of what is behind the facts in your portfolio of evidence and your technical project report. They will want to get to the bottom of why you do what you do, what the knowledge that underpins it is and where you have applied it. This will encompass the full range of KSBs and show how each of them impacts on the others and affects your ordinary working day.

On the day, it is your opportunity to clarify information and discuss your technical report and portfolio of evidence in further detail. You will be able to offer a deeper insight into the work you do and the responsibilities you carry. You will be able to show that you are not just technically able to do the job but you can also manage your time, follow programmes, follow quality controls, work in teams and take responsibility for your safety and your impact on the environment. In short, you will demonstrate that you are a rounded professional. The assessors' focus will be to confirm that you meet all of the KSBs and, if requested, confirm you are at EngTech level.

> "When I got the news I had passed, I felt a wave of relief and satisfaction that all the hard work I had done had paid off."
>
> **Ayomikun Ajayi EngTech MICE**

The overall grades available for either level 3 or level 4 are fail, pass and distinction.

Did you know?

The definitions behind the grading criteria in an EPA for both level 3 and level 4 are:

- fail: the apprentice has not met the required standards and will need to resit or retake the EPA
- pass: the apprentice has demonstrated the necessary knowledge, skills and behaviours (KSBs) to be deemed competent in their role
- distinction: the apprentice has exceeded the required standards, showing exceptional performance across all assessment components.

The EPA is your time to shine. To be at your best, visualise the day and practise. Aim for a good posture and seek to hold eye contact with the assessors. The better prepared you are, the more relaxed you will be on the day. The more relaxed you are, the more likely your assessors will be able to get to know you.

To be successful, both assessors need to be satisfied you have demonstrated all the knowledge, skills and behaviours associated with your apprenticeship.

Level 3 EPA assessment method 1: KSBs

The assessors will be looking for the following KSBs in relation to the technical project brief.

K1	Appropriate engineering principles, underpinned by appropriate mathematical, scientific and technical knowledge and understanding, relating to civil engineering and the construction process
K2	Appropriate civil engineering techniques and methods used to design, build and maintain infrastructure and buildings, the standards, contracts and specifications used, and their impact on the construction process
K3	Key principles, techniques and methods of data and technical information collection, analysis and evaluation used in delivering civil engineering models, designs, and technical solutions
K6	Industry policies, standards, regulations and codes of practice, such as Common Safety Method (CSM), Construction Design and Management (CDM) or Design Manual for Roads and Bridges (DMRB), that must be adhered to in the civil engineering environment
K7	Environmental policies and the principles of sustainable development, including those relating to the United Nations Sustainable Development Goals (SDG) and net-zero carbon emissions, and their impact on the civil engineering projects
K8	Understanding of equality, diversity and inclusion, and its impact on civil engineering solutions
K10	Methods of communication and when to use them, including how to write technical reports and using appropriate engineering terminology and conventions

S1	Apply appropriate civil engineering principles, techniques, and methods, including mathematical, scientific, and technical know-how, to civil engineering and the construction process
S2	Apply key principles, techniques and methods of data and technical information collection, analysis, and evaluation to support the delivery of civil engineering models, designs, and technical solutions
S5	Support and contribute to the production or modification of civil engineering technical solutions in accordance with relevant industry standards, regulations, and procedures and codes of practice
S6	Apply environmental policies and sustainable principles in civil engineering projects, recognising the need to reduce carbon use, lower emissions and plan for wider sustainability
S8	Consider equality, diversity and inclusion in the delivery of civil engineering projects
S10	Communicate using appropriate methods for the audience, and incorporate relevant and appropriate terms, standards, and data

B3	Applies a structured approach to problem solving with attention to detail, accuracy, and diligence

(ICE, 2024h)

Level 3 EPA assessment method 2: KSBs

The assessors will be looking for the following KSBs in relation to the portfolio of evidence.

K4	Technical drawings, modelling and designs, using computer-based software packages, such as computer aided design (CAD) or building information modelling (BIM), and their use in the sector
K5	Statutory health, safety and welfare policies, procedures, and regulations, including risk management, in relation to civil engineering project delivery
K9	Project management, quality management and assurance systems and continuous improvement as applied to civil engineering
K11	Ethical principles as applied to civil engineering and the security of data and information
K12	The values and standards by which they maintain their personal, professional and technical knowledge and skills through initial professional development (IPD) and continuing professional development (CPD)

S3	Operate appropriate software packages for data gathering and analysis, such as computer aided design (CAD) or building information modelling (BIM), to create technical drawings, models and designs using relevant conventions and engineering terminology
S4	Apply statutory health, safety and welfare policies, procedures, and regulations in the civil engineering environment, using risk management processes, procedures, and documentation
S7	Plan, carry out and manage own work in line with quality assurance, recognising the wider implications to customer needs, and within cost and resource limitations
S9	Apply document control processes and procedures using the approved processes, maintaining quality compliance when creating or amending engineering documentation
S11	Apply ethical principles to civil engineering projects, including the secure use of data and information
S12	Plan, undertake and review their own professional competence, regularly updating and reviewing their CPD to improve performance

B1	Comply with health, safety and welfare requirements, industry standards, statutory regulations, policies and codes of practice
B2	Work independently, operating in a systematic, proactive, and transparent way, using resources effectively to complete tasks, knowing their limitations and when to ask for support or escalate
B4	Is motivated when collaborating in teams, offering sensible challenge, reflects on and provides constructive feedback and contributes to discussions
B5	Maintains professional and ethical working relationships with internal, external, and connected stakeholders
B6	Takes responsibility for their own professional development, seeking opportunities to enhance their knowledge, skills, and experience

(ICE, 2024b)

Level 4 EPA assessment method 1: KSBs

The assessors will be looking for the following KSBs in relation to the technical project brief.

K1	Engineering principles, underpinned by relevant scientific, theoretical and technical knowledge and understanding to solve well-defined civil engineering problems
K2	Civil engineering techniques, procedures and methods used for civil engineering systems, to either measure and test, design, install, commission, maintain or operate
K3	Advanced mathematical, statistical and analytical problem-solving tools
K4	Properties of, and selection criteria for, materials, components or parts used in civil engineering solutions
K5	Techniques and methods to collect data and technical information, to analyse and evaluate civil engineering problems
K9	Industry policies, standards, regulations and legislation, and codes of practice, including building safety legislation, Construction (Design and Management) (CDM) or Design Manual for Roads and Bridges (DMRB)
K12	Principles of sustainable development and their impact on the lifecycle of civil engineering solutions, including United Nations Sustainable Development Goals (UNSDG), net-zero carbon emissions, environmental policies and legislations, and the climate change act
K14	Methods for planning and resourcing civil engineering tasks, and the impact on cost, quality, safety, security, and environment
K15	Methods of communication and when to use them, using appropriate engineering terminology and conventions

S1	Apply engineering principles, using relevant scientific, theoretical and technical know-how to solve well-defined civil engineering problems
S2	Apply civil engineering techniques, procedures and methods, and review the results, when measuring and testing, designing, installing, commissioning, maintaining or operating civil engineering systems
S3	Employ a range of advanced mathematical, statistical and data interpretation tools, using analytical and computational methods to interpret and solve civil engineering problems
S4	Interpret and compare performance information to choose compliant materials, components or parts
S5	Select and use technical literature and other sources of information and data to address well-defined civil engineering problems
S7	Produce civil engineering technical solutions in accordance with relevant industry standards, procedures, codes of practice, regulations, and legislation
S10	Apply principles of sustainable development, and assess the impact of these in their work
S13	Identify and use resources, equipment and technology to meet project requirements, including specifications, budget and timescales
S15	Communicate using appropriate methods for the audience, using appropriate engineering terminology and conventions
B2	Makes independent decisions when delivering civil engineering projects, while knowing their own limitations and when to ask for help or to escalate
B4	Solves problems with attention to detail, accuracy, and diligence, and seeks to continually improve

(ICE, 2024i)

Level 4 EPA assessment method 2: KSBs

The assessors will be looking for the following KSBs in relation to the portfolio of evidence.

K6	Design principles and control processes used in the civil engineering consultancy, construction or manufacturing process, and the common constraints faced
K7	Technical drawings, designs, and models, using analytical and computer-based software packages
K8	Uses and limitations of computational and digital models, including building information modelling (BIM)
K10	Statutory health, safety and welfare policies, procedures, and regulations including the Construction (Design and Management) regulation
K11	Risk assessment and mitigation processes, and their importance in the civil engineering environment
K13	Project management techniques, including quality and information management and assurance systems and continuous improvement processes
K16	Roles and responsibilities within the organisation, team dynamics and their own boundaries of authority
K17	Relationships between key organisations in the civil engineering sector (for example organisations, customers, partners and suppliers)
K18	Equality, diversity and inclusion, its importance and impact on civil engineering solutions
K19	Ethical principles as applied to civil engineering, including the need for the confidentiality and security of data and information
K20	Methods to maintain professional competence and technical knowledge, including initial professional development (IPD) and continuing professional development (CPD)

S6	Produce and interpret civil engineering technical drawings, designs, and models, using analytical and computer-based software packages, recognising the limitations of the software used
S8	Comply with, and encourage others to demonstrate, statutory health, safety and welfare policies, procedures and regulation
S9	Complete risk assessments to identify, evaluate and mitigate risks
S11	Employ project management techniques, measuring and recording progress against civil engineering project plans
S12	Assess and report on quality using appropriate management and assurance systems and continuous improvement processes
S14	Monitor and manage individual performance, and supervise others, recognising the need to comply with appropriate codes of practice and equality, diversity & inclusion (EDI) requirements

S16	Apply ethical principles to civil engineering projects, including the secure use of data and information
S17	Plan, undertake and review their own professional competence, updating and reviewing their CPD to improve performance

B1	Works to health, safety and welfare requirements, industry standards, statutory regulations and legislation, policies, and codes of practice, and ensuring others do likewise
B3	Works individually and as part of a team, being aware of their actions and the impact they may have on others, and demonstrating awareness of diversity and inclusion issues so as to meet the requirement of fairness at work
B5	Maintains professional and ethical working relationships with internal, external, and other stakeholders
B6	Takes responsibility for their own professional development, seeking opportunities to enhance their knowledge, skills, and experience, and supports others when requested

(ICE, 2024i)

Application documents comparison

Table 7.3 summarises the different application documents that should be submitted as part of the application process for level 3 and level 4 EPA.

Table 7.3 Summary of application documents

Stage 1: initial application

	EPA level 3 v1.1 (CET)	EPA level 4 v1.1 (CEST)
Application form	✓	✓
Attribute report	N/A	N/A
Portfolio of evidence (KSBs)	✓	✓
CPD	✓*	✓*
Sponsor's statement	✓*	✓*

*only applicable if applying for professional recognition as EngTech MICE

Stage 2: technical project brief (TPB)

	EPA level 3 v1.1 (CET)	EPA level 4 v1.1 (CEST)
TPB presentation	10 min	10 min
TPB report	2500 words	3500 words

The presentation and report should be combined and submitted as one pdf.

REFERENCES

DfE (Department for Education) (2022) *The Education and Skills Funding Agency (ESFA) list of acceptable qualifications for English and Maths requirements*. https://www.gov.uk/government/publications/english-and-maths-requirements-in-apprenticeship-standards-at-level-2-and-above (accessed 01/11/2024).

DfE (2024) *Apprenticeship Funding Rules*. https://assets.publishing.service.gov.uk/media/664620c4993111924d9d36ad/Apprenticeship_Funding_Rules_2024-2025_version_1.pdf (accessed 01/11/2024).

ICE (2021) *Civil Engineering Technician Apprenticeship Level 3 Mapping of Knowledge, Skills and Behaviours Against EngTech MICE Attributes, Version 2, Revision 0*. ICE, London, UK.

ICE (2024a) *Civil Engineering Senior Technician Apprenticeship Version 1.1, Version 1, Revision 3*. ICE, London, UK.

ICE (2024b) *Civil Engineering Technician (Version 1.1) End Point Assessment Application Form, Version 1, Revision 5*. ICE, London, UK.

ICE (2024c) *Level 4 Civil Engineering Senior Technician (Version 1.1) End Point Assessment Application Form, Version 1, Revision 3*. ICE, London, UK.

ICE (2024d) *Continuing Professional Development Guidance, Version 2, Revision 9*. ICE, London, UK.

ICE (2024e) *Sponsor's Statement of Support, Version 6, Revision 5*. ICE, London, UK.

ICE (2024f) *End Point Assessments Project Brief Scenarios, Version 1, Revision 4*. ICE, London, UK.

ICE (2024g) *Civil Engineering Technician Apprenticeship, Version 1.1, Version 5, Revision 6*. ICE, London, UK.

ICE (2024h) *Civil Engineering Technician Apprenticeship Level 3, Version 1.1, End Point Assessment Guidance, Version 1, Revision 3*. ICE, London, UK.

ICE (2024i) *Civil Engineering Senior Technician Apprenticeship Level 4, Version 1.1, End Point Assessment Guidance, Version 5, Revision 6*. ICE, London, UK.

IfATE (Institute for Apprenticeships and Technical Education) (2021a) *End Point Assessment Plan for Civil Engineering Technician Apprenticeship Standard*. IfATE, London, UK, licensed under the Open Government Licence. https://www.nationalarchives.gov.uk/doc/open-government-licence/version/3/

IfATE (2021b) *ST0091, Civil Engineering Technician (CET), Level 3, Version 1.1*. IfATE, London, UK.

IfATE (2022) *ST0046, Civil Engineering Senior Technician (CEST), Level 4, Version 1.1*. IfATE, London, UK.

IfATE (2024a) *ST0046, Civil Engineering Senior Technician (CEST), Level 4, Version 1.2*. IfATE, London, UK.

IfATE (2024b) *ST0046, Civil Engineering Senior Technician (CEST), Level 4, Version 1.3*. IfATE, London, UK.

Successful Professional Reviews for Civil Engineering Technicians

Malcolm Peake
ISBN 978-1-83549-943-6
https://doi.org/10.1108/978-1-83549-940-520251008
Copyright © 2025 by Malcolm Peake. Published under exclusive licence by Emerald Publishing Limited

Chapter 8
The aftermath

Once you have completed the technician professional review (TPR) or the end point assessment (EPA) you will face a lengthy wait for the result. Typically, this can be between four and six weeks. As soon as the review or EPA has been concluded, the reviewers/assessors will start to consult with each other, compare notes and discuss their observations. The assessment forms are normally completed online on the review portal and, when complete, the lead reviewer/assessor closes the form.

The staff at the ICE then know that the process is finished and they can prepare letters and take samples to put before the standards panel as part of the quality assurance process.

Did you know?

The standards panel audits a sample of the passes and every single draft failure letter. This process can take between five and ten working days. The role of the standards panel is to audit the process and check the quality of the records made by the reviewers/assessors.

All reviewers/assessors are professionally qualified members of the ICE. Once they have passed through a selection process, they undergo a rigorous training programme to prepare them for their crucial role in assessing candidates. The training ensures they are well-prepared to assess candidates effectively and uphold professional standards of the ICE. After their initial training the new reviewers/assessors undergo a series of audits while working alongside an experienced reviewer/assessor. This process fosters accountability and prepares the reviewers/assessors to work independently with confidence. After this the standards panel also conducts ongoing random checks of live TPRs and EPAs to ensure ongoing quality.

This enables you to see that everything possible is done by the ICE staff and its members to maintain the high standards of the Institution.

The successful review/assessment

Hopefully your professional TPR or EPA will be successful. Once you have achieved professional status there is no immediate target to aim for other than to progress your career. Alternatively, you may choose to follow a different path and cross over to become an engineer. This could lead to you using a sound theoretical approach to solving engineering problems that enables you to be recognised as an incorporated engineer or, if you deepen your knowledge further, to chartered engineer.

But this is not a necessity, and you may, like many technicians, prefer to build a career alongside the engineers as an independent professional technician. There are some incredibly talented people working in our industry who for one reason or another didn't go to university. However, they have bundles of invaluable knowledge learned through work on sites and in offices. There are many 'legendary' characters who are held in high regard by engineering teams for their renowned knowledge helping one way or another to save deadlines and rescue programmes from failure.

I have met and have benefited from the advice of many technician Fellows of the ICE who are not engineers and yet have a significant influence on the civil engineering industry. Several of them have contributed to this book.

Therefore, once you have been successful, you could consider becoming a mentor for the next generation. The more people who can hold up the standards of the ICE not just for themselves but share their knowledge, experience and insights with others, the better. This relationship can significantly raise standards, not just increase the number of professionally recognised technicians, but set the standard for the experience and character of those members. By fostering a culture of continuous professional development and ongoing learning, the next generation will continue to strive to be 'fit and proper people', leaving a bright future for the civil engineering industry.

The unsuccessful review/assessment

When you receive your result, if you see words such as, 'I regret to inform you…', that is just about all you will read. This will, no doubt, be a devastating blow. All the hard work and effort will feel like a waste of time. There is very little anyone can do to reduce the impact of the rejection. It is only natural to struggle to accept the reality of the situation.

Every failure letter will have been put before the standards panel. They check that the letter clearly communicates why you were unsuccessful and where you did not satisfy the reviewers. Therefore, you should have a clear explanation of what you did and did not demonstrate, but the initial shock will prevent you from looking at this letter rationally.

It is possible you will disagree with the findings. You may become disillusioned with the whole process and the ICE itself. The one thing you should never do is contact the reviewers/assessors. This will be considered highly unprofessional. The immediate response may be to appeal. This step needs to be carefully considered. The ICE allows two months for an appeal, so there is plenty of time to settle down and absorb what has happened before making any decision.

Did you know?

If the reviewers think the performance of the candidate varies significantly from the lead sponsor's comments, the ICE will get in touch to provide feedback.

Appeals procedure

The appeals procedure (ICE, 2022) is there to ensure fairness, transparency and accountability in the assessment procedures. Reasons for appeals could include administrative shortcomings related to errors in the application process, dissatisfaction with how the assessment/interview process was handled or unforeseen events that may have affected your performance during the review or EPA.

Did you know?

The *Appeals Guidance* (ICE, 2022) states:

Candidates have the right to appeal where they feel there was an error in the process, and in cases of unforeseen events. Appeals must be received within two months of the date of your result letter. Appeals after this date will not be considered.

You cannot merely say, 'I should have passed'. You need to identify where the process was not adhered to by the ICE and/or the reviewers/assessors or where there were administrative errors, procedural irregularities or disruptions in the process. If you are considering an appeal, you should first seek guidance and support from the membership support team. These are experienced members of staff; they will review with you the reasons why you were unsuccessful and recommend a course of action.

If you do appeal, then your appeal must be received by email (including payment of the fee) within two months of the date of your result notification. You should carefully and concisely lay out your grounds for appeal. The letter will be forwarded to the chair of the appeals panel and, if there are reasonable grounds for an appeal, they will contact a member of the regional support team and your lead sponsor to provide comments.

Once all the relevant documentation has been gathered, the appeals panel will assess the appeal and provide a short report addressing their findings. The appeals panel is made up of a group of highly experienced reviewers who act independently of the standards panel. They will either uphold the original decision, invite you for a resit (at no cost) or, in exceptional circumstances, overturn the original decision. However, it is highly unusual for the panel to overturn the original decision. You may be better expending your energy on getting further experience to be able to fully demonstrate the attributes at a resit.

An unsuccessful appeal will not negatively affect any future application.

Recovering from the situation

The first stage in recovering from the situation is to wait. Taking a pause can be incredibly beneficial. Pausing and reflecting will allow you to step back and give you the opportunity to assess the situation. It is essential to acknowledge the feelings that come with failure, such as anger and embarrassment. Allow yourself to feel disappointed – it's only natural to feel this way. By waiting for the initial surge of disappointment to dissipate, you will then be ready for some good advice and make some informed choices.

> ## Managing disappointment
>
> It's ok to feel disappointed; accept your feelings. Talk to someone you trust or write down your thoughts to process your emotions. Try to find something you can learn from the experience. Remember that everyone experiences setbacks. Focus on the positive. Be compassionate towards yourself. Engage in activities that you enjoy or that help you relax. Stay active; physical activity can also boost your mood. Consult with your mentors, create a plan to move forward and rectify the problem. Remember: disappointment is temporary.

The results letter usually concludes by suggesting that, before you submit again, you should seek advice, which you should certainly do. The ICE staff will be able to give you an insight into what has actually been said and help you (and quite possibly your sponsors) to 'read between the lines'. The most common phrase used is, 'You failed to demonstrate…' On the day, did you say what you meant? Or did you leave too much open for interpretation by the reviewers?

The reviewers have no hidden agenda – they are all working to the same criteria. While they may probe a particular topic, the aim is not to irritate you to such an extent it causes you to become defensive. Defensiveness hinders open communication, and this approach is therefore unlikely to allow you to present yourself favourably, so most reviewers will change the subject. The responsibility is on you to demonstrate your competence. Could you have demonstrated the required expertise? Or perhaps the reviewers have identified a real area of weakness?

Rectifying the problems

Eventually you will arrive at one of two conclusions. Either you failed to demonstrate your abilities or the reviewers have identified a weakness. If you did fail to demonstrate, then you need to take time to understand why this has happened. Had you not understood what was expected? Did the nerves get to you on the day? If you can evaluate your performance and identify what went wrong, you will be able to learn from those errors. This will put you in a stronger position and improve your chances of success when reapplying.

It is unusual, but if the reviewers have identified a weakness, then you may need to put aside personal pride. Rather than feeling defeated, view it as an opportunity to learn. Being able to bounce back from setbacks and adapt positively requires a lot of inner strength. Do not keep mulling the problem over; discuss it with as many informed people as possible. Any shortfall can be put right with further experience. Focus on the lessons and use them to enhance your skills.

You need to talk the problem through with key people in your organisation, such as your sponsors or line managers. They should be able to identify opportunities for suitable experience to improve your abilities. This way you will be able to move forwards. This approach should benefit your career as well as your chances of passing the review next time.

The revised submission

If, or when, you choose to reapply, you will need to look at your entire application and see whether there are any adjustments you could make to improve any perceived weaknesses. Your sponsors must submit new statements of support. During this time, you are likely to have gained further experiences, which you may choose to include in your next application. For example, as a minimum you should refresh and update your CPD records.

Do not be tempted to think that you only need to fix a couple of problems. You are taking the whole review again. On the next occasion you must demonstrate all the Attributes and satisfy all the ICE requirements. With a little reflection and the right help and support, you should be able to prepare for success the next time you apply.

REFERENCE

ICE (2022) *Appeals Guidance, Version 2, Revision 4.* ICE, London, UK.

Malcolm Peake
ISBN 978-1-83549-943-6
https://doi.org/10.1108/978-1-83549-940-520251009

Index

www.ingramcontent.com/pod-product-compliance
Lightning Source LLC
Chambersburg PA
CBHW060313220326
41598CB00027B/4316